普通高等学校计算机类专业特色教材·精选系列

J2EE 项目开发与设计

（第二版）

主　编　彭灿华　韦晓敏　魏士伟
副主编　崔建明　杨呈永　郎佳南　陈玲萍

中国铁道出版社有限公司
CHINA RAILWAY PUBLISHING HOUSE CO., LTD.

内 容 简 介

本书共分为 13 章，包括三篇：基础篇、提高篇、综合篇。"基础篇"从环境部署、开发工具的安装讲起，由浅入深，详细讲述使用 MySQL 作为后台数据库进行 J2EE（已更名为 Java EE）项目开发的方法，同时结合每一章的知识点讲述相关实例，加深对知识点的理解。"提高篇"分别讲述 Java EE 黄金组合 Struts2 与 Hibernate，并结合实例详细说明各框架的使用。"综合篇"详细讲述 Spring 的入门、操作与使用，以及 Spring、Struts2 和 Hibernate 这三个框架的整合原理和步骤，最后以一个 Java EE 项目对前面介绍的三个框架的相关知识点和内容进行回顾和复习。

本书是将理论知识运用到实际开发中的实践和尝试，适合作为高等院校计算机相关专业的教材，也可作为社会培训班的教材及软件设计人员的辅导用书。

图书在版编目（CIP）数据

J2EE 项目开发与设计 / 彭灿华，韦晓敏，魏士伟主编. — 2 版. — 北京：中国铁道出版社，2016.8（2022.12重印）

普通高等学校计算机类专业特色教材. 精选系列

ISBN 978-7-113-22156-0

Ⅰ. ①J… Ⅱ. ①彭… ②韦… ③魏… Ⅲ. ①JAVA 语言－程序设计－高等学校－教材 Ⅳ. ①TP312.8

中国版本图书馆 CIP 数据核字（2016）第 206494 号

书　　名：	J2EE 项目开发与设计
作　　者：	彭灿华　韦晓敏　魏士伟

策　　划：	祝和谊	编辑部电话：	（010）63549508
责任编辑：	周　欣　徐盼欣		
封面设计：	一克米工作室		
封面制作：	白　雪		
责任校对：	王　杰		
责任印制：	樊启鹏		

出版发行：	中国铁道出版社有限公司（100054，北京市西城区右安门西街8号）
网　　址：	http://www.tdpress.com/51eds
印　　刷：	三河市航远印刷有限公司
版　　次：	2013 年 2 月第 1 版　2016 年 8 月第 2 版　2022 年 12 月第 9 次印刷
开　　本：	787 mm×1 092 mm　1/16　印张：20.5　字数：480 千
书　　号：	ISBN 978-7-113-22156-0
定　　价：	48.00 元

版权所有　侵权必究

凡购买铁道版图书，如有印制质量问题，请与本社教材图书营销部联系调换。电话：（010）63550836

打击盗版举报电话：（010）63549461

第二版前言

本书内容

首先感谢各位读者对本书第一版提出的宝贵建议,经过一段时间的修改,《J2EE 项目开发与设计(第二版)》终于面世了。第二版在保留第一版全部优点和特色的基础上,作了部分优化、改进和创新。这些优化、改进和创新的最终目的是用最浅显易懂的案例和教学流程帮助软件设计人员快速掌握 Web 开发技术的使用,并能将其应用在实战中。

本书共分为 13 章,包括三篇:基础篇、提高篇、综合篇。"基础篇"从环境部署、开发工具的安装讲起,由浅入深,详细讲述使用 MySQL 作为后台数据库进行 J2EE(已更名为 Java EE)项目开发的方法,同时结合每一章的知识点讲述相关实例,加深对知识点的理解。"提高篇"分别讲述 Java EE 黄金组合 Struts2 与 Hibernate,并结合实例详细说明各框架的使用。"综合篇"详细讲述 Spring 的入门、操作与使用,以及 Spring、Struts2 和 Hibernate 这三个框架的整合原理和步骤,最后以一个 Java EE 项目对前面介绍的三个框架的相关知识点和内容进行回顾和复习。严格按照软件工程的规范,详细讲述项目的背景与目标、需求分析、总体设计、数据库设计、详细设计与实现,使读者深刻体会项目开发的各个环节,提升综合开发能力和实际动手能力。项目中使用了基于 Struts2+Hibernate+Spring 框架的开发方式,通过学习,读者可以对 Java EE 开发技术有更加深入和透彻的理解。

本书由桂林理工大学彭灿华、桂林电子科技大学信息科技学院韦晓敏、桂林航天工业学院魏士伟任主编,负责拟定编写大纲,组织协调并总纂定稿。桂林理工大学崔建明、杨呈永,桂林理工大学南宁分校电气与电子工程系郎佳南,桂林电子科技大学信息科技学院陈玲萍任副主编。具体分工如下:第 1 章至第 4 章由桂林电子科技大学信息科技学院韦晓敏编写,第 5 章由桂林理工大学南宁分校电气与电子工程系郎佳南编写,第 6 章由桂林理工大学崔建明、杨呈永编写,第 7 章由桂林理工大学彭灿华编写,第 8 章由桂林电子科技大学信息科技学院陈玲萍编写,第 9 章、第 10 章主要由桂林航天工业学院魏士伟和彭灿华编写,第 11 章至第 13 章由魏士伟编写。

本书特色

本书是将理论知识运用到实际开发中的实践和尝试,详细列出每个案例的开发步骤,实例易于阅读和理解。综合案例以软件工程的标准设计并开发，编写理念面向需求、面向市场。

本书适用对象

本书适合作为高等院校计算机相关专业的教材，也可作为社会培训班的教材及软件设计人员的辅导用书。

由于水平有限，书中疏漏之处在所难免，恳请读者批评指正。读者如果有好的意见与建议或者在学习的过程中遇到不解的地方，可以通过邮件进行探讨。

联系方法如下：

电子邮箱：449271349@qq.com

索取本书源代码及各章节调试视频可发邮件与编者联系。

编　者

2016 年 6 月

第一版前言

本书内容

本书是笔者在多年项目开发过程中的经验总结，本书通过丰富的实例，由浅入深、循序渐进地介绍了目前采用 Java 进行 Web 开发的各种框架的使用方法，从而帮助软件设计人员快速掌握 Web 开发技术，并能将其应用在实战中。

本书共分为 13 章，包括三大部分：基础篇、提高篇、综合篇。"基础篇"从环境部署、开发工具的安装讲起，由浅入深，详细讲述使用 MySQL 作为后台数据库进行 J2EE 项目开发的方法，同时结合每章的知识点讲述相关实例，加深对知识点的理解。"提高篇"分别讲述 J2EE 黄金组合 Struts2、Hibernate 与 Spring 框架，并结合实例详细说明各框架的使用。"综合篇"严格按照软件工程的规范，详细讲述项目的背景与目标、需求分析、总体设计、数据库设计、详细设计与实现，使读者深刻体会项目开发的各个环节，提升综合开发能力和实际动手能力。案例中使用了基于 Struts2+Hibernate+Spring 框架的开发方式，通过学习，可以对 J2EE 开发技术有更加深入和透彻的理解。

本书由彭灿华、魏士伟担任主编，负责拟定编写大纲，组织协调并定稿。由韦晓敏、张振华、吉伟明、莫岚、廖建锋、于彬任副主编。具体编写分工如下：第 1 章、第 2 章由桂林电子科技大学信息科技学院韦晓敏与张振华编写；第 3 章、第 4 章由桂林电子科技大学信息科技学院彭灿华与河南经贸职业学院廖建锋编写；第 5 章由南阳理工学院于彬编写；第 6 章、第 7 章由彭灿华编写；第 8 章至第 10 章主要由桂林航天工业学院魏士伟和桂林电子科技大学信息科技学院彭灿华编写；第 11 章至第 13 章由魏士伟和福建农林大学计算机学院吉伟明编写。另外，感谢桂林医学院莫岚、桂林电子科技大学信息科技学院杨呈永、郭建、陈玲萍、雷光圣、宋若翔，桂林电子科技大学陈金龙、向荣，天津天狮学院袁英与吕向风对全书统稿工作进行的指导，感谢资深程序员蓝周磊给予的悉心指导。

本书特色

本书详细列出每个案例的开发步骤，实例易于阅读和理解。综合案例以软件工程的标准设计并开发，使读者的编程理念面向需求、面向市场。

本书适用对象

本书既可作为高等院校相关专业的教材或教学参考书，也可作为社会培训班的教材及初级、中级、高级软件设计人员的辅导书。

由于水平有限，书中错漏之处在所难免，恳请读者批评指正。读者如果有好的意见与建议或者在学习的过程中遇到不解的地方，可以通过邮件进行探讨。

联系方法如下：

电子邮箱：449271349@qq.com

网址：www.pengcanhua.cn。本书源代码及相关视频录像可以在该网站免费下载。

编　者

2012 年 10 月

目　录

基　础　篇

第1章　开发环境的搭建 1
1.1　Java EE 开发环境所需软件 1
1.2　Java EE 开发环境配置 3
1.3　MyEclipse 汉化与优化 9
　　1.3.1　MyEclipse 汉化 9
　　1.3.2　MyEclipse 优化 12
1.4　使用 MyEclipse 创建和发布 Web Project 13
　　1.4.1　编写输出 HelloWorld 的 JSP 文件 13
　　1.4.2　运行 JSP 文件 14
小结 16
习题 16

第2章　JSP 技术详解 17
2.1　JSP 技术简介 17
2.2　JSP 的基本语法 18
　　2.2.1　基本语句 21
　　2.2.2　数据类型 22
2.3　JSP 的内置对象 23
　　2.3.1　request 对象 23
　　2.3.2　response 对象 28
　　2.3.3　pageContext 对象 30
　　2.3.4　session 对象 30
　　2.3.5　application 对象 30
　　2.3.6　out 对象 31
2.4　JSP 技术应用——登录功能 32
小结 35
习题 35

第3章　Servlet 技术详解 37
3.1　Servlet 技术简介 37
　　3.1.1　Servlet 的概念 37
　　3.1.2　Servlet 的生命周期 38

3.1.3 Servlet 的重要函数 38
3.1.4 开发第一个 Servlet 39
3.2 站点计数监听器制作 47
小结 50
习题 50

第 4 章 JSP 中使用 JavaBean 51

4.1 JavaBean 简介 51
4.1.1 JavaBean 的属性 52
4.1.2 JavaBean 的方法 53
4.2 创建一个 JavaBean 53
4.3 在 JSP 中调用 JavaBean 57
4.4 JavaDoc 文档的生成 60
4.5 JAR 插件的制作与使用 62
4.5.1 JAR 相关特点 62
4.5.2 JAR 的使用 63
小结 64
习题 65

第 5 章 搭建数据库开发环境 67

5.1 MySQL 概述 67
5.1.1 MySQL 简介 67
5.1.2 下载并安装 MySQL 68
5.1.3 下载并安装 Navicat for MySQL 72
5.2 使用 MySQL 数据库 73
5.2.1 采用 Navicat 管理 MySQL 数据库 73
5.2.2 创建数据库 74
5.2.3 创建数据表 74
5.2.4 新增记录 75
5.3 SQL 语法介绍 75
5.3.1 SQL 简介 75
5.3.2 SQL 基本语法 75
小结 77
习题 77

第 6 章 JDBC 技术详解 79

6.1 JDBC 概述 79
6.2 JDBC 数据库连接 80
6.2.1 连接 MySQL 数据库 80

  6.2.2 连接 SQL Server 2000 数据库 85
  6.2.3 连接 SQL Server 2005 数据库 87
 6.3 JSP 操作 MySQL 数据库 90
  6.3.1 数据查询 91
  6.3.2 数据添加 100
  6.3.3 数据编辑 104
  6.3.4 数据删除 108
 小结 109
 习题 109

第 7 章 综合实例——BLOG 系统开发 110

 7.1 功能要求 110
 7.2 数据库设计 110
  7.2.1 数据库的需求分析 110
  7.2.2 数据库的逻辑设计 111
 7.3 框架搭建 112
 7.4 功能实现 119
  7.4.1 通用功能实现 119
  7.4.2 数据访问层功能实现 129
  7.4.3 后台表示层功能实现 139
  7.4.4 前台表示层功能实现 153
 7.5 系统运行界面 159
  7.5.1 前台界面 159
  7.5.2 后台界面 160
 小结 161
 习题 162

提 高 篇

第 8 章 Struts2 入门 163

 8.1 Struts2 框架介绍 163
  8.1.1 Struts1 概述 163
  8.1.2 MVC 概述 164
  8.1.3 WebWork 概述 165
  8.1.4 Struts2 概述及优势 165
 8.2 Struts2 的环境配置 165
  8.2.1 下载 Struts2 框架包 165
  8.2.2 搭建 Struts2 开发环境 166
 8.3 第一个 Struts2 示例 167

8.3.1　准备工作 168
　　8.3.2　配置 struts.xml 与 struts.properties 文件 168
　　8.3.3　创建控制器（Action 类） 169
　　8.3.4　创建视图层 170
　　8.3.5　测试运行 171
小结 172
习题 172

第 9 章　Struts2 框架技术 173

9.1　Struts2 的标签库 173
　　9.1.1　Struts2 标签库的使用 174
　　9.1.2　if/elseif/else 标签 174
　　9.1.3　iterator 标签 175
　　9.1.4　include 标签 178
　　9.1.5　property 标签 179
　　9.1.6　部分 UI 标签的使用 179
9.2　Struts2 的国际化操作 181
　　9.2.1　Struts2 实现国际化的原理 181
　　9.2.2　实现国际化步骤 182
9.3　Struts2 数据验证 187
　　9.3.1　使用 validate()方法进行验证 187
　　9.3.2　使用配置文件进行验证 191
小结 194
习题 194

第 10 章　Hibernate 概述及实例分析 195

10.1　Hibernate 框架介绍 195
　　10.1.1　持久化和 ORM 简介 195
　　10.1.2　Hibernate 框架 196
10.2　Hibernate 的环境配置 196
　　10.2.1　下载 Hibernate 框架包 196
　　10.2.2　搭建 Hibernate 开发环境 197
10.3　第一个 Hibernate 示例 203
　　10.3.1　准备工作 203
　　10.3.2　创建 POJO 和 Hibernate 映射文件 204
　　10.3.3　修改 Hibernate 配置文件 207
　　10.3.4　创建操作数据库的主类：NewsOperator 208
　　10.3.5　数据查询 210
　　10.3.6　数据编辑 212

10.3.7 数据删除 213
10.3.8 测试 215
小结 215
习题 216

综 合 篇

第 11 章 Spring 入门 217
11.1 Spring 框架介绍 217
11.2 Spring 的环境配置 218
    11.2.1 下载 Spring 框架包 218
    11.2.2 搭建 Spring 开发环境 219
11.3 第一个 Spring 示例 224
    11.3.1 准备工作 224
    11.3.2 编写接口文件 225
    11.3.3 编写实现接口文件 225
    11.3.4 修改 Spring 的配置文件 applicationContext.xml 226
    11.3.5 创建调用组件的主程序类 226
    11.3.6 测试运行 227
小结 228
习题 228

第 12 章 使用 Spring 操作数据库 229
12.1 数据源 datasource 的注入 229
12.2 Spring 框架的事务处理 230
    12.2.1 传统的 JDBC 事务处理 230
    12.2.2 Spring 框架的事务处理 231
12.3 PlatformTransactionManager 的接口作用 233
12.4 使用 Template 访问数据 237
    12.4.1 Template 模式简介 237
    12.4.2 HibernateTemplate 的使用 238
小结 241
习题 242

第 13 章 Spring+Struts2+Hibernate 集成实例 243
13.1 项目需求 243
    13.1.1 项目需求概述 243
    13.1.2 系统框架 244
13.2 数据库的设计 245

13.3 配置开发环境 251
13.3.1 web.xml 文件的配置 251
13.3.2 Spring 配置文件 applicationContext.xml 的配置 252
13.3.3 Struts2 配置文件 Struts.xml 的配置 256
13.3.4 国际化资源文件的配置 260
13.4 编写持久化对象（PO） 260
13.4.1 定义 Book 类及映射文件 261
13.4.2 定义 Bargain 类及映射文件 266
13.4.3 定义 Orders 类及映射文件 267
13.4.4 定义 Ordersbook 类及映射文件 269
13.4.5 定义 User 类及映射文件 271
13.5 建立数据库访问层组件（DAO） 274
13.5.1 DAO 组件接口的定义 274
13.5.2 实现 DAO 组件 277
13.5.3 配置 DAO 组件 285
13.6 创建业务层组件 286
13.6.1 业务逻辑组件接口的定义 286
13.6.2 实现业务逻辑组件 289
13.6.3 事务管理配置 294
13.6.4 配置业务逻辑组件 294
13.7 创建业务控制器 295
13.7.1 业务控制器的执行流程 295
13.7.2 网上书店系统 Action 类分析 295
13.8 创建视图 JSP 页面 297
13.8.1 用户注册界面 297
13.8.2 用户登录界面 298
13.8.3 用户信息修改界面 298
13.8.4 系统首页界面 299
13.8.5 显示图书详细信息界面 301
13.8.6 购物车界面 302
13.8.7 显示用户订单列表界面 304
13.8.8 添加图书界面 306
13.9 运行网上书店系统 307
13.9.1 系统前台界面 307
13.9.2 系统后台界面 309
小结 310
习题 311
附录 A 常见数据类型转换 312
参考文献 316

基 础 篇

第 1 章
开发环境的搭建

学习目标
- 了解 Java EE 运行环境配置。
- 掌握 MyEclipse、Tomcat、JDK 的安装与集成开发。
- 了解 MyEclipse 开发工具的使用与优化。
- 制作第一个 Web 应用程序。

本书介绍的 Java EE 编程都将使用本章搭建的开发环境。为提高开发效率，所涉及软件均与 MyEclipse 进行了整合。

1.1 Java EE 开发环境所需软件

本书所选用开发环境的所需软件清单如表 1-1 所示。

表 1-1 所需软件清单

软件名称	版 本 号	说 明	下 载 地 址
MyEclipse	2015	myeclipse-2015-offline-installer-windows	https://downloads.myeclipseide.com/downloads/products/eworkbench/2015/installers/myeclipse-2015-2014-07-11-offline-installer-windows.exe
Tomcat	8.0	apache-tomcat-8.0.33	http://tomcat.apache.org/download-80.cgi
JDK	8	jdk-8u91-windows-x64	http://www.oracle.com/technetwork/java/javase/downloads/index-jsp-138363.html
操作系统	Windows 7/10	32 位/64 位	

> **注意**
> 本书使用的操作系统为 64 位，如果所用操作系统为 32 位，则可在官方网站下载相应的软件。

为帮助读者更准确地下载合适的软件，下面给出本书所有软件的下载官方网址。

（1）Tomcat 官方网站：http://tomcat.apache.org/。其下载页面如图 1-1 所示，选择版本页面如图 1-2 所示。

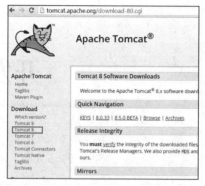

图 1-1　Tomcat 下载页面

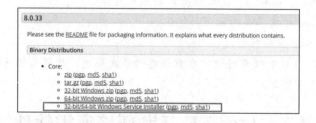

图 1-2　选择 Tomcat 版本页面

（2）JDK 官方网站：http://www.oracle.com。其下载页面如图 1-3 所示，选择版本页面如图 1-4 所示。

图 1-3　JDK 下载页面

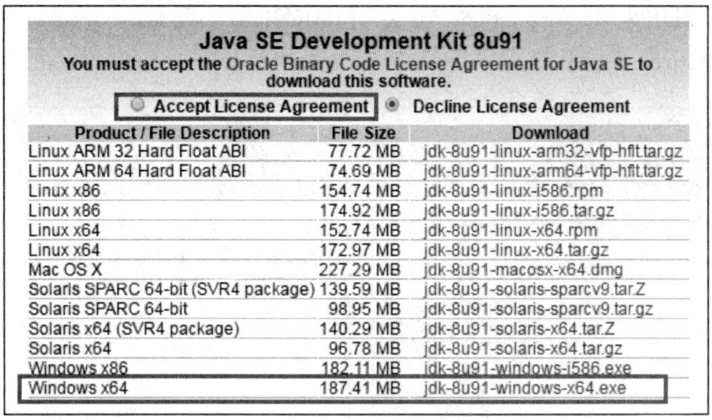

第 1 章 开发环境的搭建

图 1-4 选择 JDK 版本页面

（3）MyEclipse 官方网站：http://www.myeclipsecn.com。MyEclipse 2015 下载地址：http://www.myeclipsecn.com/download/。其下载页面如图 1-5 所示。

图 1-5 MyEclipse 2015 下载页面

1.2 Java EE 开发环境配置

第一步：安装 JDK8（默认安装路径为：C:\Program Files\Java\jdk1.8.0_91\），其安装过程如图 1-6～图 1-10 所示。

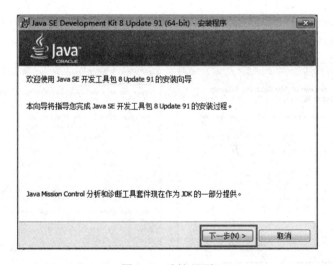

图 1-6 欢迎界面

-3-

在安装 JDK8 时，会同时安装好 JRE1.8（默认安装路径为：C:\Program Files\Java\jre1.8.0_91）。

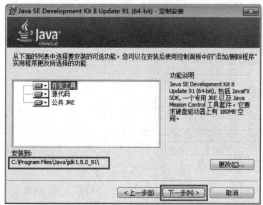

图 1-7 "定制安装"界面　　　　　图 1-8 "进度"界面

图 1-9 "目标文件夹"界面　　　　图 1-10 "完成"界面

第二步：安装 Tomcat 8.0 （默认安装路径为：C:\Program Files\Apache Software Foundation\Tomcat 8.0），如图 1-11～图 1-18 所示。

 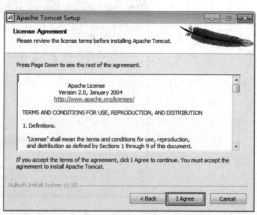

图 1-11 Tomcat 安装程序欢迎界面　　图 1-12 许可协议界面

第 1 章 开发环境的搭建

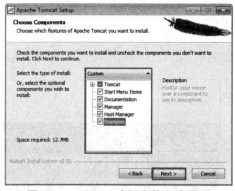

图 1-13　Tomcat 选择安装方式界面

图 1-14　端口配置界面

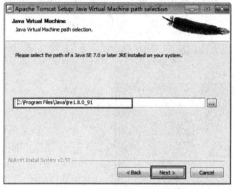

图 1-15　默认选择 JRE 路径界面

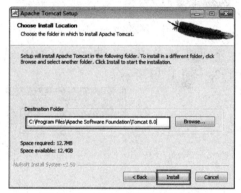

图 1-16　Tomcat 路径设置界面

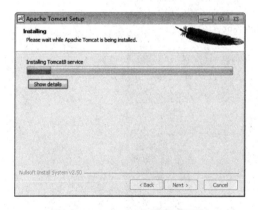

图 1-17　Tomcat 安装进度界面

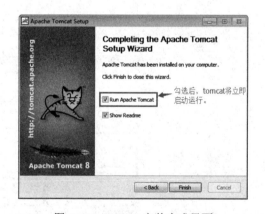

图 1-18　Tomcat 安装完成界面

启动 Tomcat 服务，如图 1-19 所示。Tomcat 服务状态图标如图 1-20 所示。

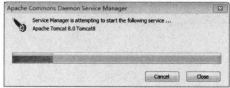

图 1-19　启动 Tomcat 服务

图 1-20　Tomcat 服务状态图标

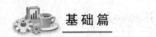

第三步：安装 MyEclipse 2015。其安装界面如图 1-21 所示，启动界面如图 1-22 所示。

图 1-21　MyEclipse 2015 安装界面　　　　图 1-22　MyEclipse 2015 启动界面

下面将 MyEclipse、Tomcat、JDK 进行整合，以提高开发效率。具体操作步骤如图 1-23～图 1-30 所示。

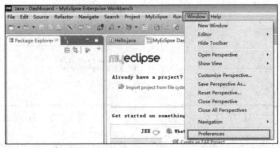

图 1-23　选择 Preferences 命令

图 1-24　打开 New Server Runtime Environment 对话框

第1章 开发环境的搭建

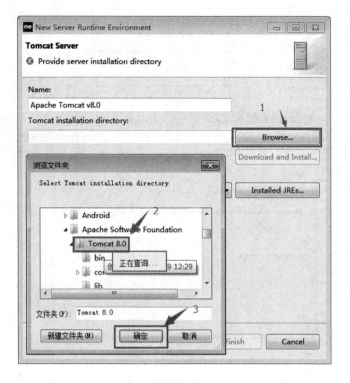

图 1-25　选择 Tomcat 8.0

图 1-26　设置安装目录

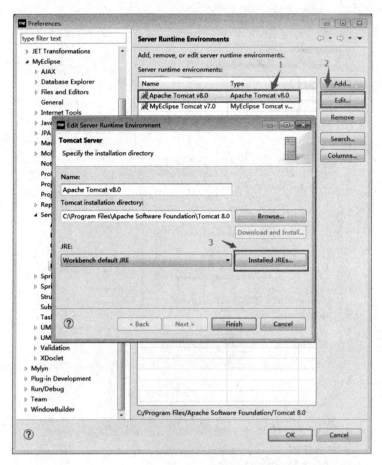

图 1-27 打开 Edit Server Runtime Environment 对话框

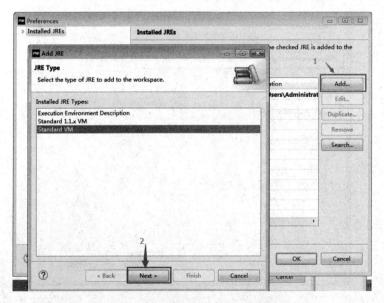

图 1-28 打开 Add JRE 对话框

第 1 章 开发环境的搭建

图 1-29 选择 JRE

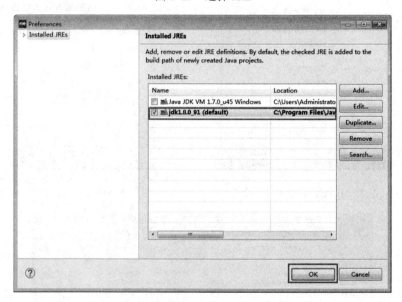

图 1-30 完成整合

至此，MyEclipse、Tomcat 与 JDK 三个软件实现了整合，为以后开发基于 Java EE 的软件提供了开发环境。

1.3 MyEclipse 汉化与优化

1.3.1 MyEclipse 汉化

第一步：下载 MyEclipse 2015 汉化包。该汉化包可从网上搜索并下载到，搜索关键词为

-9-

"myeclispe 2015 汉化"。解压此汉化包，目录结构如图 1-31 所示。

图 1-31 MyEclipse 汉化包目录结构

第二步：复制 MyEclipse 安装目录路径，如图 1-32～图 1-34 所示。

图 1-32 选择 MyEclipse 2015 C1 命令

图 1-33 单击"打开文件位置"按钮

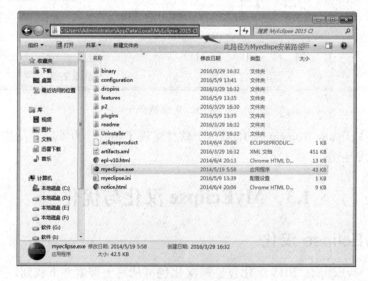

图 1-34 复制安装目录路径

第 1 章 开发环境的搭建

第三步：双击汉化包中的 MyEclipse2015HH.bat 文件。Common 文件中的内容如图 1-35 所示。

图 1-35　Common 文件中的内容

第四步：新建 Java 项目，如图 1-36 和图 1-37 所示。

图 1-36　新建 Java 项目

第五步：保存，重启 MyEclipse，即可看到中文界面。

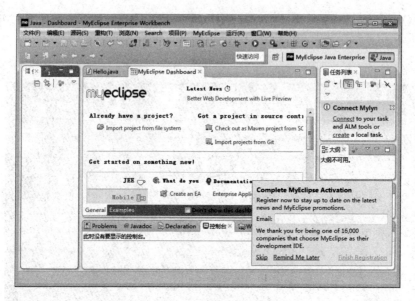

图 1-37　建好的 Java 项目

1.3.2　MyEclipse 优化

在"窗口"→"首选项"→"Java"→"编辑器"→"内容辅助"中添加"Java 的自动激活触发器"字符，如图 1-38 所示，启用代码提醒功能。

图 1-38　代码提醒功能设置

1.4 使用 MyEclipse 创建和发布 Web Project

1.4.1 编写输出 HelloWorld 的 JSP 文件

创建 HelloWorld 工程，如图 1-39~图 1-43 所示。

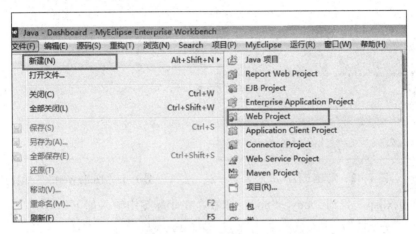

图 1-39 选择"新建"→Web Project 命令

图 1-40 新建 Web 工程　　　　　　　图 1-41 配置工程

图 1-42 配置工程库

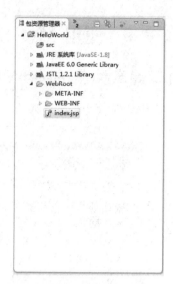

图 1-43 HelloWorld 工程目录结构

双击 index.jsp，找到<body></body>标签，在此标签中输入如下代码。

```
<%
 out.println("Hello World!");
%>
```

1.4.2 运行 JSP 文件

第一步：部署 HelloWorld 工程，如图 1-44～图 1-46 所示。

图 1-44 单击"部署"按钮

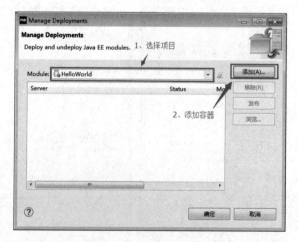

图 1-45 管理部署

第 1 章 开发环境的搭建

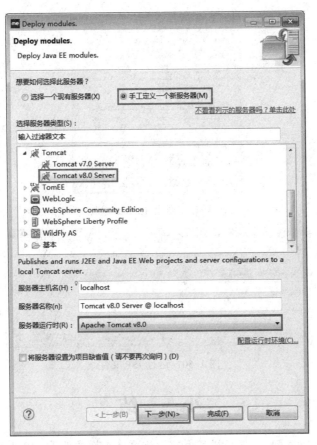

图 1-46 部署模块

第二步：启动 Tomcat 8，如图 1-47 所示。提示端口占用错误，如图 1-48 所示。

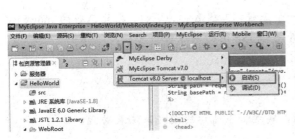

图 1-47 启动 Tomcat 8

图 1-48 端口占用错误提示

上述问题的解决办法是结束 Tomcat 进程，如图 1-49 所示。

再次在 MyEclipse 中启动 Tomcat，问题得到解决。

第三步：打开浏览器，在地址栏中输入 http://localhost:8080/HelloWorld/，运行 JSP 文件，如图 1-50 所示。

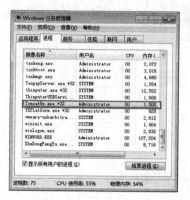

图 1-49　结束 Tomcat 进程

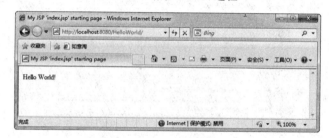

图 1-50　运行 JSP 文件

小　　结

本章着重介绍 Java EE 开发环境的搭建，本章 1.4 节提供了一个 JSP 开发的例子，通过学习这个例子，读者应该对 JSP 开发有一个最初的认识。

习　　题

1. 配置好 Java EE 开发环境，新建一个工程，工程名称为 Ex1，在此工程的 index.jsp 文件中调试如下代码，查看运行结果，并总结 JSP 运行的原理。

```
<%
for (int i=0;i<10;i++){
    out.print("this is "+i);
}
%>
```

2. 在工程 Ex1 中新建一个 JSP 页面，实现如下所示的九九乘法表功能。

```
1*1=1
1*2=2  2*2=4
1*3=3  2*3=6  3*3=9
1*4=4  2*4=8  3*4=12 4*4=16
1*5=5  2*5=10 3*5=15 4*5=20 5*5=25
1*6=6  2*6=12 3*6=18 4*6=24 5*6=30 6*6=36
1*7=7  2*7=14 3*7=21 4*7=28 5*7=35 6*7=42 7*7=49
1*8=8  2*8=16 3*8=24 4*8=32 5*8=40 6*8=48 7*8=56 8*8=64
1*9=9  2*9=18 3*9=27 4*9=36 5*9=45 6*9=54 7*9=63 8*9=72 9*9=81
```

第 2 章
JSP 技术详解

学习目标
- 掌握 JSP 的基本语法：JSP 注释、JSP 脚本元素、JSP 指令、JSP 动作指令。
- 掌握 JSP 内建对象。
- 熟练使用 JSP 的内置对象在开发中的应用。

本章介绍了 JSP 的基本语法，为简化 JSP 开发提供了一些内建对象，讲述了 JSP 的内建对象及案例分析。首先讲述了 JSP 内建对象的基本概念，然后分别详细介绍了内建对象：request、response、pageContext、session、application、out、config、page 的概念和使用。

2.1 JSP 技术简介

JSP（JavaServer Pages）是由 Sun 公司倡导、许多公司参与建立的一种动态网页技术标准。JSP 技术类似 ASP 技术，它在传统的网页 HTML 文件（*.htm,*.html）中插入 Java 程序段（Scriptlet）和 JSP 标记（tag），从而形成 JSP 文件（*.jsp）。

用 JSP 开发的 Web 应用是跨平台的，既能在 Linux 下运行，也能在其他操作系统上运行。

JSP 技术使用 Java 编程语言编写类 XML 的 tags 和 scriptlets 封装产生动态网页的处理逻辑。网页还能通过 tags 和 scriptlets 访问存在于服务端的资源的应用逻辑。JSP 将网页逻辑与网页设计和显示分离，支持可重用的基于组件的设计，使基于 Web 的应用程序开发变得迅速和容易。

Web 服务器在遇到访问 JSP 网页的请求时，首先执行其中的程序段，然后将执行结果连同 JSP 文件中的 HTML 代码一起返回给客户。插入的 Java 程序段可以操作数据库、重新定向网页等，以实现建立动态网页所需要的功能。

JSP 与 Java Servlet 一样，是在服务器端执行的，通常返回该客户端的就是一个 HTML 文本，因此客户端只要有浏览器就能浏览。

JSP 的 1.0 规范的最后版本是 1999 年 9 月推出的，同年 12 月又推出了 1.1 规范。目前较新的是 JSP2.0 规范。

JSP 页面由 HTML 代码和嵌入其中的 Java 代码所组成。服务器在页面被客户端请求

以后对这些Java代码进行处理，然后将生成的HTML页面返回给客户端的浏览器。Java Servlet是JSP的技术基础，而且大型的Web应用程序的开发需要Java Servlet和JSP配合才能完成。JSP具备Java技术的简单易用、完全的面向对象、具有平台无关性且安全可靠、主要面向因特网等所有特点。

1. JSP技术的强势

（1）一次编写，到处运行。在这一点上Java比PHP更出色，除了系统之外，代码不用做任何更改。

（2）系统的多平台支持。基本上可以在所有平台上的任意环境中开发，在任意环境中进行系统部署，在任意环境中扩展。

（3）强大的可伸缩性。从只有一个小的Jar文件就可以运行Servlet/JSP，到由多台服务器进行集群和负载均衡，到多台Application进行事务处理、消息处理，从一台服务器到无数台服务器，Java显示出了巨大的生命力。

（4）多样化和功能强大的开发工具支持。这一点与ASP很像，Java已经有了许多非常优秀的开发工具，而且许多可以免费得到，并且其中许多已经可以顺利地运行于多种平台之下。

2. JSP技术的弱势

（1）与ASP一样，Java的一些优势正是它致命的问题所在。正是由于为了跨平台的功能和极度的伸缩能力，所以极大地增加了产品的复杂性。

（2）Java的运行速度是用class常驻内存来完成的，所以它在一些情况下所使用的内存比起用户数量来说确实是"最低性能价格比"了。另一方面，它还需要硬盘空间来存储一系列的.java文件和.class文件，以及对应的版本文件。

2.2 JSP的基本语法

本节主要介绍JSP页面的基本结构，让读者对JSP页面有一个全面的了解。例2-1是一个简单的JSP文件。

例2-1　计算1+2+…+100的和。

```
<%@ page language="java" import="java.util.*" pageEncoding="utf-8"%>
<!-- 这是一个简单的实例 -->
<!DOCTYPE HTML PUBLIC "-//W3C//DTD HTML 4.01 Transitional//EN">
<html>
    <head>
        <title>计算1+2+...+100的和</title>
    </head>
    <body>
    <%
    int m=100,n=1;
    int sum=0;
    while(n<=m){            /*判断循环条件*/
        sum=sum+n;
```

```
        n++;
    }
%>
<%--此处输出 sum 的值 --%>
<%=sum %>
</body>
</html>
```

1．JSP 注释

为提高程序代码的可读性，会在编写代码时加上必要的注释，用于标注程序开发过程的开发提示。

（1）HTML 注释。该注释主要用于在客户端动态地显示一个注释，也称输出注释。

语法格式：`<!–注释内容-->`

（2）JSP 注释。该注释隐藏在 JSP 源代码中，它不会输出到客户端，也称隐藏注释。

语法格式：`<%--注释内容--%>`

（3）scriptlets 中的注释。通常使用"//"表示单行注释，使用"/**   */"表示多行注释。

在 IE 中通过"查看"菜单中的"查看源文件"命令可查看注释在客户端的运行情况。

2．JSP 表达式

JSP 提供了一种输出表达式值的简单方法，输出表达式的语法格式如下：

`<%=表达式%>`

使用输出表达式的语法代替了原来的 out.println 输出语句，但是输出表达式后没有分号。

3．常用编译指令

JSP 的编译指令是通知 JSP 引擎的消息，它不直接生成输出。编译指令有默认值，因此开发人员无须为每个指令设置值。

（1）page 指令。page 指令是针对当前页面的指令，它作用于整个 JSP 页面。它通常位于 JSP 页面的顶端，一个 JSP 页面可以有多条 page 指令，但是其属性只能出现一次，重复的属性设置将覆盖先前的设置。

语法格式：

```
<%@ page
    [language="java"]
    [import="{package.class|package.*},..."]
    [contentType="TYPE;charset=CHARSET"]
    [session="true|false"]
    [buffer="none|8kb|sizekb"]
    [autoFlash="true|false"]
    [isThreadSafe="true|false"]
    [info="text"]
    [errorPage="relativeURL"
```

```
        [isErrorPage="true|false"]
        [extends="package.class"]
        [isELIgnored="true|false"]
        [pageEncoding="CHARSET"]
%>
```

language：定义要使用的脚本语言，目前只能是"java"，即 language="java"。

import：和一般的 Java import 意义一样，用于引入要使用的类，用逗号","隔开包或者类列表。默认省略，即不引入其他类或者包，默认导入的包有：java.lang.*;、javax.servlet.* 、 javax.servlet.jsp.* 和 javax.servlet.http.* 。 导入包格式如下：import="java.io.*,java.util.Hashtable"。

session：指定所在页面是否使用 session 对象。默认值为 true，即 session="true"。

buffer：指定到客户输出流的缓冲模式。如果为 none，则不缓冲；如果指定数值，那么输出就用不小于这个值的缓冲区进行缓冲，单位为 KB。它与 autoFlash 一起使用。默认不小于 8 KB，根据不同的服务器可设置。例如，buffer="64kb"。

autoFlash：如果为 true，缓冲区满时，到客户端输出被刷新；如果为 false，缓冲区满时，出现运行异常，表示缓冲区溢出。默认为 true，即 autoFlash="true"。

info：关于 JSP 页面的信息，定义一个字符串，可以使用 servlet.getServletInfo()获得。默认省略。例如，info="测试页面"。

isErrorPage：表明当前页是否为其他页的 errorPage 目标。如果被设置为 true，则可以使用 exception 对象；如果被设置为 false，则不可以使用 exception 对象。例如，isErrorPage="true"。默认为 false。

errorPage：定义此页面出现异常时调用的页面。默认忽略，例如 errorPage="error.jsp"。

isThreadSafe：用来设置 JSP 文件是否能多线程使用。如果设置为 true，那么一个 JSP 能够同时处理多个用户的请求；如果设置为 false，那么一个 JSP 只能一次处理一个请求。例如，isThreadSafe="true"。

contentType：定义 JSP 字符编码和页面响应的 MIME 类型。TYPE=MIME TYPE;charset=CHARSET。默认为 TYPE=text/html,CHARSET=iso8859-1。例如，contentType="text/html;charset=gb2312"。注意：如果 charset 指令没有设置为 gb2312 或 UTF-8，JSP 程序中的中文将显示乱码。

pageEncoding：JSP 页面的字符编码，默认值为 pageEncoding="iso-8859-1"。例如，pageEncoding="gb2312"。

isELIgnored：指定 EL（表达式语言）是否被忽略。如果为 true，则容器忽略"${}"表达式的计算。默认值由 web.xml 描述文件的版本确定，Servlet2.3 以前的版本将忽略。例如 isELIgnored="true"。

（2）include 指令。include 指令通知容器将当前 JSP 页面中的内嵌的、在指定位置上的资源内容包含。被包含的文件可以被 JSP 解析。这种解析发生在编译期间。

include 指令的格式：

```
<%@include file="包含文件名称"%>
```

由于使用了 include 指令，可以把一个复杂的 JSP 页面分成若干简单部分，这样可以大大提高 JSP 页面的管理的效率，当要对页面进行更改时，只需要更改对应的部分就可以了。

例 2-2 include 的使用。创建一个页面命名为 index.jsp，该页面中包含四个文件，如图 2-1 所示。

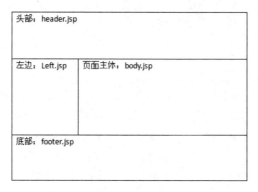

图 2-1 include 的使用效果

> **注 意**
> （1）被 include 包含文件中最好去掉<html>、<body>等标签，否则页面排版不正确。
> （2）避免被包含和包含文件中定义相同的变量和方法。

2.2.1 基本语句

1. 条件语句

（1）条件语句 if 有三种使用形式：单分支 if 语句、if...else 语句和多分支 if 语句。

单分支 if 语句格式如下：
```
if(表达式)
    语句1;
```

if...else 语句格式如下：
```
if(表达式)
    语句1;
else
    语句2;
```

多分支 if 语句格式如下：
```
if(表达式1)
    语句1;
else if(表达式2)
    语句2;
else if(表达式3)
    语句3;
    …
else
    语句n;
```

（2）条件语句 switch。switch 语句格式如下：
```
switch(表达式)
{
    case 常量表达式1:
        语句1;
        break;
    case 常量表达式2:
        语句2;
        break;
        …
    case 常量表达式n:
```

```
        语句 n;
        break;
    default:
        语句 m;
        break;
}
```

2. 循环语句

（1）当型循环。while 循环是当型循环，其语法格式如下：

```
while(表达式)
    循环体;
```

（2）do...while 循环。do while 循环语句的语法格式如下：

```
do {
    循环体;
}while(表达式);
```

（3）for 循环。for 循环的语法格式如下：

```
for(表达式1;表达式 2;表达式 3)
    循环体;
```

2.2.2 数据类型

基本数据类型有以下四种：

（1）int 长度数据类型有：byte（8 bit）、short（16 bit）、int（32 bit）、long（64 bit）。

（2）float 长度数据类型有：单精度（32 bit float）、双精度（64 bit double）。

（3）boolean 类型变量的取值有：true、false。

（4）char 数据类型有：unicode 字符，16 bit。

字符串转化为基本数据类型的说明及示例代码如表 2-1 所示。

表 2-1　字符串转化为基本数据类型的说明与示例代码

说　　明	代　　码
将一个字符串转化成一个 Integer 对象	int i=Integer.valueOf("123").intValue()
将一个字符串转化成一个 Float 对象	float f=Float.valueOf("123").floatValue()
将一个字符串转化成一个 Boolean 对象	boolean b=Boolean.valueOf("123").booleanValue()
将一个字符串转化成一个 Double 对象	double d=Double.valueOf("123").doubleValue()
将一个字符串转化成一个 Long 对象	long l=Long.valueOf("123").longValue()
将一个字符串转化成一个 Character 对象	char=Character.valueOf("123").charValue()

从低精度向高精度转换顺序：

byte→short→int→long→float→double→char。

> **注　意**
> 两个 char 型数据运算时，自动转换为 int 型；当 char 型与其他类型运算时，也会先自动转换为 int 型数据，再做其他类型的自动转换。

例 2-3 将字符串转成整型。代码如下：

```
public static int toInt(String input){
    try{
        return Integer.parseInt(input);
    }catch(Exception e){
        return 0;
    }
}
```

在进行类型转换时，一般会将这些方法写在一个 JavaBean 中（后面会作详细介绍），以方便调用。

2.3　JSP 的内置对象

2.3.1　request 对象

request 封装了用户提交给服务器的所有信息，通过 request 对象的方法可以获取用户提交的信息。用户通常使用 HTML 中的 Form 表单向服务器的某个 JSP 或者 Servlet 提交信息。代码如下：

```
<form name="表单名字" method= "get | post" action= "提交信息的目标地址">
    提交内容
</form>
```

> **注意**
> （1）get 方法和 post 方法的主要区别：使用 get 方法提交的信息会在提交的过程中显示在浏览器的地址栏中；而使用 post 方法提交的信息不会显示在地址栏中。
> （2）提交内容的方式：文本框、密码框、下拉列表、单选按钮、多选按钮、文本区域等。

1．获取用户提交信息

request 对象可以使用 getParameter(string s)方法获取该表单通过 text 提交的信息。如：request.getParameter("boy")。

下面通过一个实例解释如何进行表单数据的处理。

例 2-4 表单数据的处理。

第一步：先创建一个 one.jsp 文件，代码如下：

one.jsp 代码：

```
<%@ page contentType="text/html; charset=utf-8" language="java" import="java.util.*" errorPage="" %>
<!DOCTYPE html PUBLIC "-//W3C//DTD XHTML 1.0 Transitional//EN" "http://www.w3.org/TR/xhtml1/DTD/xhtml1-transitional.dtd">
<html xmlns="http://www.w3.org/1999/xhtml">
<head>
<meta http-equiv="Content-Type" content="text/html; charset=utf-8" />
<title>request 对象使用-one.jsp</title>
```

```
</head>
<body>
<form action="two.jsp" method="post" name="form">
    <input type="text" name="boy">
    <input type="submit" value="Enter" name="submit">
</form >
</body>
</html>
```

第二步：创建一个页面 two.jsp，用于接收 one.jsp 提交的表单数据。

two.jsp 代码：

```
<%@ page contentType="text/html; charset=utf-8" language="java"
import="java.util.*" errorPage="" %>
<!DOCTYPE html PUBLIC "-//W3C//DTD XHTML 1.0 Transitional//EN"
"http://www.w3.org/TR/xhtml1/DTD/xhtml1-transitional.dtd">
<html xmlns="http://www.w3.org/1999/xhtml">
<head>
<meta http-equiv="Content-Type" content="text/html; charset=utf-8" />
<title>request 对象使用-two.jsp</title>
</head>

<body>
获取文本框提交的信息：
<%String textContent=request.getParameter("boy"); %>
<%=textContent%>
</body>
</html>
```

> **注意**
>
> 使用 request 对象获取信息要格外小心，要避免使用空对象，否则会出现 NullPointerException 异常，所以可以作以下处理。
>
> ```
> <%
>     String textContent=request.getParameter("boy");
>     if(textContent != null && !"".equals(textContent))/*判断是否为空情况*/
>     {
>        out.print(textContent);
>     }
>     else
>     {
>        out.print("textContent 为 null");
>     }
> %>
> ```

2. 处理汉字信息

当 request 对象获取客户提交的汉字字符时，会出现乱码问题，必须进行特殊处理。

可以使用 request 对象的 setCharacterEncoding()方法进行编码转换。具体代码如下：

```jsp
<%@ page language="java" import="java.util.*" pageEncoding="utf-8"%>
<%
    request.setCharacterEncoding("utf-8");   //解决中文乱码
%>
<!DOCTYPE html PUBLIC "-//W3C//DTD XHTML 1.0 Transitional//EN"
"http://www.w3.org/TR/xhtml1/DTD/xhtml1-transitional.dtd">
<html xmlns="http://www.w3.org/1999/xhtml">
<head>
    <meta http-equiv="Content-Type" content="text/html; charset=utf-8" />
    <title>request 对象使用-two.jsp</title>
</head>

<body>
    获取文本框提交的信息:
    <%
    String textContent = request.getParameter("boy");
    if(textContent != null && !"".equals(textContent))  /*判断是否为空情况*/
    {
        out.print(textContent);
    } else {
        out.print("textContent 为 null");
    }
    %>
</body>
</html>
```

常用方法：
- getProtocol()：获取客户向服务器提交信息所使用的通信协议，比如 HTTP/1.1 等。
- getServletPath()：获取客户请求的 JSP 页面文件的目录。
- getContentLength()：获取客户提交的整个信息的长度。
- getMethod()：获取客户提交信息的方式，比如 post 或 get。
- getHeader(String s)：获取 HTTP 头文件中由参数 s 指定的头名字的值。一般来说，s 参数可取的头名有：accept、referer、accept-language、content-type、accept-encoding、user-agent、host、content-length、connection、cookie 等，比如，s 取值 user-agent 将获取客户的浏览器的版本号等信息。
- getHeaderNames()：获取头名字的一个枚举。
- getHeaders(String s)：获取头文件中指定头名字的全部值的一个枚举。
- getRemoteAddr()：获取客户的 IP 地址。
- getRemoteHost()：获取客户机的名称（如果获取不到，就获取 IP 地址）。
- getServerName()：获取服务器的名称。
- getServerPort()：获取服务器的端口号。
- getParameterNames()：获取客户提交的信息体部分中 name 参数值的一个枚举。

request 对象使用：

```jsp
<%@ page contentType="text/html; charset=utf-8" language="java" import="java.util.*" errorPage="" %>
<!DOCTYPE html PUBLIC "-//W3C//DTD XHTML 1.0 Transitional//EN" "http://www.w3.org/TR/xhtml1/DTD/xhtml1-transitional.dtd">
<html xmlns="http://www.w3.org/1999/xhtml">
<head>
    <meta http-equiv="Content-Type" content="text/html; charset=utf-8" />
    <title>request 对象</title>
</head>
<body>
    <BR>客户使用的协议是:
        <% String protocol=request.getProtocol();
           out.println(protocol);
        %>
    <BR>获取接受客户提交信息的页面:
        <% String path=request.getServletPath();
           out.println(path);
        %>
    <BR>接受客户提交信息的长度:
        <% int length=request.getContentLength();
           out.println(length);
        %>
    <BR>客户提交信息的方式:
        <% String method=request.getMethod();
           out.println(method);
        %>
    <BR>获取HTTP头文件中User-Agent的值:
        <% String header1=request.getHeader("User-Agent");
           out.println(header1);
        %>
    <BR>获取HTTP头文件中accept的值:
        <% String header2=request.getHeader("accept");
           out.println(header2);
        %>
    <BR>获取HTTP头文件中Host的值:
        <% String header3=request.getHeader("Host");
           out.println(header3);
        %>
    <BR>获取HTTP头文件中accept-encoding的值:
        <% String header4=request.getHeader("accept-encoding");
           out.println(header4);
        %>
    <BR>获取客户的IP地址:
        <% String  IP=request.getRemoteAddr();
           out.println(IP);
        %>
    <BR>获取客户机的名称:
        <% String clientName=request.getRemoteHost();
```

```
        out.println(clientName);
    %>
<BR>获取服务器的名称:
    <% String serverName=request.getServerName();
        out.println(serverName);
    %>
<BR>获取服务器的端口号:
    <% int serverPort=request.getServerPort();
        out.println(serverPort);
    %>
</body>
</html>
```

例 2-5   request 对象使用实例。实现一个类似在线考试的问卷提交功能。
three.jsp 代码：

```
<%@ page contentType="text/html; charset=utf-8" language="java" import="java.util.*" errorPage="" %>
<!DOCTYPE html PUBLIC "-//W3C//DTD XHTML 1.0 Transitional//EN" "http://www.w3.org/TR/xhtml1/DTD/xhtml1-transitional.dtd">
<html xmlns="http://www.w3.org/1999/xhtml">
<head>
<meta http-equiv="Content-Type" content="text/html; charset=utf-8" />
<title>three.jsp</title>
</head>
<body>
    <P>
    <FORM action="four.jsp" method="post" name="form1">
    诗人李白是中国历史上哪个朝代的人: <BR>
        <INPUT type="radio" name="R" value="a">宋朝
        <INPUT type="radio" name="R" value="b">唐朝
        <INPUT type="radio" name="R" value="c">明朝
        <INPUT type="radio" name="R" value="d" >元朝
         <BR>
    <P>小说《红楼梦》的作者是:
         <BR>
        <INPUT type="radio" name="P" value="a">曹雪芹
        <INPUT type="radio" name="P" value="b">罗贯中
        <INPUT type="radio" name="P" value="c">李白
        <INPUT type="radio" name="P" value="d">司马迁
        <BR>
        <INPUT TYPE="submit" value="提交答案" name="submit">
    </FORM>
</body>
</html>
```

four.jsp 代码：

```
<%@ page contentType="text/html; charset=utf-8" language="java" import="java.util.*" errorPage="" %>
<!DOCTYPE html PUBLIC "-//W3C//DTD XHTML 1.0 Transitional//EN" "http://www.w3.org/TR/xhtml1/DTD/xhtml1-transitional.dtd">
```

```
<html xmlns="http://www.w3.org/1999/xhtml">
<head>
  <meta http-equiv="Content-Type" content="text/html; charset=utf-8" />
  <title>four.jsp</title>
</head>
<body>
    <% int n=0;
      String s1=request.getParameter("R");
      String s2=request.getParameter("P");
      if(s1==null)
      {s1="";}
      if(s2==null)
      {s2="";}
      if(s1.equals("b"))
      { n++;}
      if(s2.equals("a"))
      { n++;}
    %>
    <P>您得了<%=n%>分
</body>
</html>
```

2.3.2 response 对象

对客户的请求做出动态的响应，向客户端发送数据。在某些情况下，当响应客户时，需要将客户重新引导至另一个页面，可以使用 response 的 sendRedirect(URL)方法实现客户的重定向。新建一个文件 goto.jsp，代码如下所示。

```
<%@ page contentType="text/html; charset=utf-8" language="java"
import="java.util.*" errorPage="" %>
<!DOCTYPE html PUBLIC "-//W3C//DTD XHTML 1.0 Transitional//EN"
"http://www.w3.org/TR/xhtml1/DTD/xhtml1-transitional.dtd">
<html xmlns="http://www.w3.org/1999/xhtml">
<head>
  <meta http-equiv="Content-Type" content="text/html; charset=utf-8" />
  <title>goto.jsp</title>
</head>
<body>
    <%
    String address = request.getParameter("where");
    if(address!=null){
       if(address.equals("Baidu"))
          response.sendRedirect("http://www.baidu.com");
        else if(address.equals("Yahoo"))
          response.sendRedirect("http://www.yahoo.com");
        else if(address.equals("Sun"))
          response.sendRedirect("http://www.sun.com");
    }
    %>
```

```
    <b>Please select:</b><br>
    <form action="goto.jsp" method="post">
    <select name="where">
       <option value="Baidu" selected>go to Baidu
       <option value="Yahoo" > go to Yahoo
       <option value="Sun" > go to Sun
    </select>
    <input type="submit" value="go">
    </form>
</body>
</html>
```

response 对象是一个 javax.servlet.http.HttpServletResponse 类的子类的对象。response 常用方法举例：

- addCookie(Cookie cook)：添加一个 cookie 对象，用来保存客户端的用户信息。
- addheader(String name,String value)：添加 HTTP 文件头信息。
- constrainsHeader(String name)：判断指定名字的 HTTP 文件头是否已经存在，返回一个布尔值。
- encodeURL()：使用 sessionId 封装 URL。如果没有必要封装 URL，则返回原值。
- flushBuffer()：强制把当前缓冲区的内容发送到客户端。
- getBufferSize()：返回缓冲区的大小。
- getOutputStream()：返回到客户端的输出流对象。
- sendError(int)：向客户端发送错误的信息。例如，404 是指网页不存在或者请求页面无效。
- sendRedirect(String url)：把响应发送到另一个位置进行处理。
- setContentType(String contentType)：设置响应的 MIME 类型。
- setHeader(String name,String value)：设置指定名字的 HTTP 文件头的值。

setHeader 页面自动跳转：

```
<body>
  <%
  response.setHeader("refresh","3;URL=common.jsp?ref=aaa") ;
  %>
  三秒后跳转!!<br>
  如果没有跳转,请按<a href="common.jsp">这里</a>!!!
</body>
```

setHeader 禁用页面缓存：在实际的项目开发工程中，往往会用到禁用缓存技术，即如果通过后退按钮回到了某一页，也必须从服务器上重新读取。例如：

```
<%@page contentType="text/html;charset=utf-8" import="java.util.Date"%>
<html>
<head>
  <%
  response.setHeader("Cache-Control","no-cache");
  response.setHeader("Pragma","no-cache");
  response.setDateHeader ("Expires", 0);
```

```
    %>
    <title>禁用页面缓存</title>
</head>
<body>
    <%
        Date d = new Date();
        System.out.println(d.toLocaleString());
    %>
</body>
</html>
```

运行修改过后的这个例子，就会发现当单击"后退"按钮回到 index.jsp 时，页面代码都会被执行一次。

2.3.3　pageContext 对象

pageContext 是一个页面上下文对象。JSP 引入了一个名为 pageContext 的类，通过它可以访问页面的许多属性。

pageContext 类拥有 getRequest()、getResponse()、getOut()、getSession()等方法。

pageContext 变量存储与当前页面相关联的 PageContext 对象的值。

如果方法需要访问多个与页面相关的对象，传递 pageContext 要比传递 request、response、out 等的独立引用更容易。（虽然两种方式都能达到同样的目的。）

2.3.4　session 对象

session 对象指的是客户端与服务器的一次会话，从客户端连到服务器的一个 WebApplication 开始，直到客户端与服务器断开连接为止。它是 HttpSession 类的实例。

常用方法：

- long getCreationTime()：返回 session 创建时间。
- public String getId()：返回 session 创建时 JSP 引擎为它设的唯一 ID 号。
- long getLastAccessedTime()：返回此 session 里客户端最近一次请求时间。
- int getMaxInactiveInterval()：返回两次请求间隔多长时间此 session 被取消（ms）。
- String[] getValueNames()：返回一个包含此 session 中所有可用属性的数组。
- void invalidate()：取消 session，使 session 不可用。
- boolean isNew()：返回服务器创建的一个 session，客户端是否已经加入。
- void removeValue(String name)：删除 session 中指定的属性。
- void setMaxInactiveInterval()：设置两次请求间隔多长时间此 session 被取消（ms）。

2.3.5　application 对象

application 对象实现了用户间数据的共享，可存放全局变量。它开始于服务器的启动，终止于服务器的关闭，在此期间，此对象将一直存在。在用户的前后连接或不同用户之间的连接中，可以对此对象的同一属性进行操作。在任何地方，对此对象属性的操作都将影响其他用户对此的访问。服务器的启动和关闭决定了 application 对象的生命。它是 ServletContext 类的实例。

常用方法：
- Object getAttribute(String name)：返回给定名的属性值。
- Enumeration getAttributeNames()：返回所有可用属性名的枚举。
- void setAttribute(String name,Object obj)：设定属性的属性值。
- void removeAttribute(String name)：删除一属性及其属性值。
- String getServerInfo()：返回 JSP(servlet)引擎名及版本号。
- String getRealPath(String path)：返回一虚拟路径的真实路径。
- ServletContext getContext(String uripath)：返回指定 WebApplication 的 Application 对象。
- int getMajorVersion()：返回服务器支持的 Servlet API 的最大版本号。
- int getMinorVersion()：返回服务器支持的 Servlet API 的最小版本号。
- String getMimeType(String file)：返回指定文件的 MIME 类型。
- URL getResource(String path)：返回指定资源（文件及目录）的 URL 路径。
- InputStream getResourceAsStream(String path)：返回指定资源的输入流。
- RequestDispatcher getRequestDispatcher(String uripath)：返回指定资源的 RequestDispatcher 对象。
- Servlet getServlet(String name)：返回指定名的 Servlet。
- Enumeration getServlets()：返回所有 Servlet 的枚举。
- Enumeration getServletNames()：返回所有 Servlet 名的枚举。
- void log(String msg)：把指定消息写入 Servlet 的日志文件。
- void log(Exception exception,String msg)：把指定异常的栈轨迹及错误消息写入 Servlet 的日志文件。
- 19 void log(String msg,Throwable throwable)：把栈轨迹及给出的 Throwable 异常的说明信息写入 Servlet 的日志文件。

2.3.6　out 对象

out 对象是一个输出流，用来向客户端输出数据。out 对象用于各种数据的输出。
常用方法：
- out.print()：输出各种类型数据。
- out.newLine()：输出一个换行符。
- out.close()：关闭流。

out 对象的示例代码如下：

```
<%@ page contentType="text/html; charset=utf-8" language="java"
import="java.util.*" errorPage="" %>
<!DOCTYPE html PUBLIC "-//W3C//DTD XHTML 1.0 Transitional//EN"
"http://www.w3.org/TR/xhtml1/DTD/xhtml1-transitional.dtd">
<html xmlns="http://www.w3.org/1999/xhtml">
<head>
    <meta http-equiv="Content-Type" content="text/html; charset=utf-8" />
    <title>three.jsp</title>
</head>
```

```
<body>
<%
Date Now = new Date();
String hours=String.valueOf(Now.getHours());
String mins=String.valueOf(Now.getMinutes());
String secs=String.valueOf(Now.getSeconds());
%>
现在是
<%out.print(String.valueOf(Now.getHours()));%>
小时
<%out.print(String.valueOf(Now.getMinutes()));%>
分
<%out.print(String.valueOf(Now.getSeconds()));%>
秒
</body>
</html>
```

2.4 JSP 技术应用——登录功能

本实例框架如图 2-2 所示。其首页是 index.jsp，它包括顶部的菜单栏 menu.jsp，左边的登录栏 login.jsp，右边的主页 main.jsp。

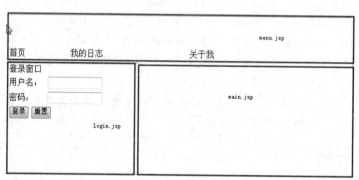

图 2-2 登录前的效果

第一步：启动 MyEclipse，新建一个名称为 ch2-login 的 Web 工程，目录结构图如图 2-3 所示。

图 2-3 ch2-login 工程目录结构图

第二步：编辑 menu.jsp 文件，代码如下所示。

```jsp
<%@ page language="java" import="java.util.*" pageEncoding="utf-8"%>
<table width="525" border="0">
 <tr>
    <td>首页</td>
    <td>我的日志</td>
    <td>关于我</td>
 </tr>
</table>
```

第三步：编辑 main.jsp 文件，代码如下所示。

```jsp
<%@ page language="java" import="java.util.*" pageEncoding="utf-8"%>
<!-- 此文件暂无其他内容 -->
```

第四步：编辑 login.jsp 文件，代码如下所示。

```jsp
<%@ page language="java" import="java.util.*" pageEncoding="utf-8"%>
<%
if("LoginAction".equals(request.getParameter("action")))
{
    String txtUsername=request.getParameter("txtUsername");
    String txtPassword=request.getParameter("txtPassword");
    //假设正确的用户名与密码是 admin ,1234
    //接下来判断用户输入的用户与密码是否正确
    if("admin".equals(txtUsername) && "1234".equals(txtPassword))
    {
        session.setAttribute("CurrentUser",txtUsername);
        response.sendRedirect("index.jsp");
    }
    else
    {
        out.print("<script>alert('用户名或密码不正确');window.loaction.href('index.jsp');</script>");
    }
}
%>
<%if( session.getValue("CurrentUser")!=null) { %>
<table width="200" border="0">
 <tr>
    <td>欢迎您：<%=session.getValue("CurrentUser") %></td>
 </tr>
 <tr>
    <td><a href="edit.jsp">编辑个人资料</a></td>
 </tr>
```

```jsp
    <tr>
      <td><a href="logout.jsp">退出登录</a></td>
    </tr>
  </table>
<%}else{ %>
<form id="form1" name="form1" method="post" action="login.jsp?action=LoginAction">
   <table width="199" border="0">
    <tr>
      <td colspan="2">登录窗口</td>
    </tr>
    <tr>
      <td>用户名: </td>
      <td><input name="txtUsername" type="text" size="12" /></td>
    </tr>
    <tr>
      <td>密码: </td>
      <td><input name="txtPassword" type="text" size="12" /></td>
    </tr>
    <tr>
      <td colspan="2"><input type="submit" name="button" id="button" value="登录" />
        <input type="reset" name="button2" id="button2" value="重置" /></td>
    </tr>
   </table>
 </form>
<%} %>
```

第五步: 编辑 index.jsp 文件, 代码如下所示。

```jsp
<%@ page language="java" import="java.util.*" pageEncoding="utf-8"%>
<!DOCTYPE HTML PUBLIC "-//W3C//DTD HTML 4.01 Transitional//EN">
<html>
  <head>
    <title>index.jsp</title>
  </head>
  <body>
    <table width="673" height="301" border="1">
    <tr>
      <td height="87" colspan="2" valign="bottom">
       <!-- 包含menu.jsp页到index.jsp文件中 -->
       <%@ include file="menu.jsp"%>
      </td>
    </tr>
    <tr>
      <td width="231" valign="top">
       <!-- 包含login.jsp页到index.jsp文件中 -->
```

```
        <%@include file="login.jsp"%>
      </td>
      <td width="372">
        <!-- 包含main.jsp页到index.jsp文件中 -->
        <%@include file="main.jsp"%>
      </td>
    </tr>
  </table>
</body>
</html>
```

第六步：新建 logout.jsp 文件，实现退出，代码如下所示。

```
<%@ page language="java" import="java.util.*" pageEncoding="utf-8"%>
<%
    session.invalidate();
    response.sendRedirect("index.jsp");
%>
```

第七步：部署工程，并测试运行实例，如图 2-4 所示。

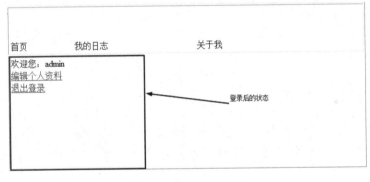

图 2-4　登录后的效果

小　　结

本章主要介绍了 JSP 的内建对象的相关概念，并且为每个内建对象提供了使用实例。

JSP 中的内建对象有 request、response、pageContext、session、application、out、config、page、exception。

out 对象是 JSP 开发中使用最频繁，也是 JSP 初学者每个需要掌握的内建对象之一，主要作用是输出动态内容。作为初学者，应重点掌握 out、request、response、session 的使用。本章最后结合一个实际开发中的案例将 JSP 的内建对象使用作了详细介绍。

习　　题

1. 配置好 Java EE 开发环境，新建一个工程，工程名称为 Ex2，在此工程中新建一个 reg.jsp 文件，界面效果如图 2-5 所示。

图 2-5　reg.jsp 文件效果

2. 新建一个接收注册信息的页面 save.jsp，将 reg.jsp 页面中的信息打印输出（注意中文乱码问题的解决）。

第 3 章
Servlet 技术详解

学习目标
- Servlet 相关概念介绍。
- Servlet 常用的类和接口介绍。
- 使用 HttpServlet 处理客户端请求。
- 使用 Servlet 生成动态图像。

本章首先介绍讲述了 Servlet 技术的工作原理，接着介绍了 JSP 与 Servlet 之间的关系、Servlet 的应用范围及其缺陷、Servlet 的生命周期，以及 Servlet 常用类、接口使用。通过本章学习，读者能够掌握 Servlet 编程技术。

3.1 Servlet 技术简介

3.1.1 Servlet 的概念

要在网上浏览网页，需要一个 Web 服务器，浏览网页的过程就是浏览器通过 HTTP 协议与 Web 服务器交互的过程。在过去，大多是静态网页，因此只须把资源放在 Web 服务器上即可。如今随着应用的发展，客户机与服务器需要动态的交互，为了实现这一目的，需要开发一个遵循 HTTP 协议的服务器端应用软件，来处理各种请求。Servlet 是一个基于 Java 技术的 Web 组件，运行在服务器端，利用 Servlet 可以很轻松地扩展 Web 服务器的功能，使它满足特定的应用需要。Servlet 由 Servlet 容器管理，Servlet 容器也叫 Servlet 引擎，是 Servlet 的运行环境，给发送的请求和响应提供网络服务。比如，Tomcat 就是常用的一个 Servlet 容器，接受客户机请求并做出响应的步骤如下：

（1）客户机访问 Web 服务器，发送 HTTP 请求。
（2）Web 服务器接收到请求后，传递给 Servlet 容器。
（3）Servlet 容器加载 Servlet，产生 Servlet 实例，并向其传递表示请求和响应的对象。
（4）Servlet 得到客户机的请求信息，并进行相应的处理。
（5）Servlet 实例把处理结果发送回客户机，容器负责确保响应正确送出，同时将控制返回给 Web 服务器，如图 3-1 所示。

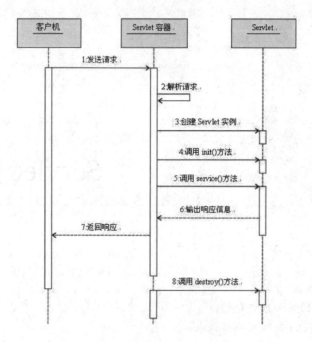

图 3-1　Servlet 对客户机提供服务的过程

3.1.2　Servlet 的生命周期

Servlet 部署在容器里（如 Tomcat、JBoss 等），它的生命周期由容器来管理。Servlet 的生命周期概括为以下几个阶段：

（1）装载 Servlet，此操作一般为动态执行。有些服务器提供相应的管理功能，可以在启动的时候就装载 Servlet 并能够初始化特定的 Servlet。

（2）创建一个 Servlet 实例。

（3）调用 Servlet 的 init()方法。

（4）服务，如果容器接收到对此 Servlet 的请求，那么它调用 Servlet 的 service()方法。

（5）销毁，通过调用 Servlet 的 destory()方法来销毁 Servlet。

在以上几个阶段中，第（3）阶段的提供服务是最重要的阶段。

3.1.3　Servlet 的重要函数

HttpServlet 是 GenericServlet 的一个派生类，为基于 HTTP 协议的 Servlet 提供了基本的支持；HttpServlet 类包含 init()、destory()、Service()等方法。其中 init()和 destory()方法是继承的。

1．init()方法

在 Servlet 的生命期中，仅执行一次 init()方法。它是在服务器装入 Servlet 时执行的。可以配置服务器，以在启动服务器或客户机首次访问 Servlet 时装入 Servlet。无论有多少客户机访问 Servlet，都不会重复执行 init()。

默认的 init()方法通常是符合要求的，但也可以定制 init()方法来覆盖它，典型的是管理服务器端资源。例如，初始化数据库连接等。因此，所有覆盖 init()方法的 Servlet 应调用

super.init()以确保仍然执行这些任务。在调用 service()方法之前，须确保已完成 init()方法。

2．service()方法

service() 方法是 Servlet 的核心。每当一个客户请求一个 HttpServlet 对象，该对象的 service() 方法就要被调用，而且传递给这个方法一个"请求"（ServletRequest）对象和一个"响应"（ServletResponse）对象作为参数。在 HttpServlet 中已存在 service() 方法。默认的服务功能是调用与 HTTP 请求的方法相应的 do 功能。例如，如果 HTTP 请求方法为 GET，则默认情况下调用 doGet()。Servlet 应该为 Servlet 支持的 HTTP 方法覆盖 do 功能。因为 HttpServlet.service() 方法会检查请求方法是否调用了适当的处理方法，不必要覆盖 service() 方法，只需覆盖相应的 do 方法就可以了。

Servlet 的响应可以是下列几种类型：

（1）一个输出流，浏览器根据它的内容类型（如 text/HTML）进行解释。
（2）一个 HTTP 错误响应，重定向到另一个 URL、Servlet、JSP。

3．destory()方法

destroy() 方法仅执行一次，即在服务器停止且卸载 Servlet 时执行该方法。可以将 Servlet 作为服务器进程的一部分来关闭。默认的 destroy() 方法通常是符合要求的，但也可以覆盖它，典型的是管理服务器端资源。例如，如果 Servlet 在运行时会累计统计数据，则可以编写一个 destroy() 方法，该方法用于在未载入 Servlet 时将统计数字保存在文件中。另一个示例是关闭数据库连接。

当服务器卸载 Servlet 时，将在所有 service() 方法调用完成后，或在指定的时间间隔过后调用 destroy() 方法。一个 Servlet 在运行 service() 方法时可能会产生其他的线程，因此须确认在调用 destroy() 方法时，这些线程已终止或完成。

4．getServletConfig()方法

getServletConfig()方法返回一个 ServletConfig 对象，该对象用来返回初始化参数和 ServletContext。ServletContext 接口提供有关 Servlet 的环境信息。

5．getServletInfo()方法

getServletInfo()方法是一个可选的方法，它提供有关 Servlet 的信息，如作者、版本、版权。

当服务器调用 Servlet 的 service()、doGet()和 doPost()这三个方法时，均需要"请求"和"响应"对象作为参数。"请求"对象提供有关请求的信息，而"响应"对象提供一个将响应信息返回给浏览器的通信途径。

javax.servlet 软件包中的相关类为 ServletResponse 和 ServletRequest，而 javax.servlet.http 软件包中的相关类为 HttpServletRequest 和 HttpServletResponse。Servlet 通过这些对象与服务器通信并最终与客户机通信。Servlet 能通过调用"请求"对象的方法获知客户机环境，服务器环境的信息和所有由客户机提供的信息。Servlet 可以调用"响应"对象的方法发送响应，该响应是准备发回客户机的。

3.1.4 开发第一个 Servlet

下面来开发第一个简单的 Servlet。

第一步：启动 MyEclipse，新建一个 Web 工程，如图 3-2 所示。

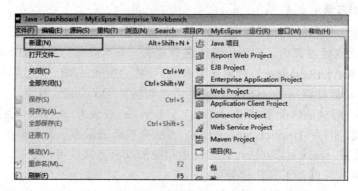

图 3-2 新建一个 Web 工程

在弹出的对话框中输入工程名称，并设置工程项目保存路径，如图 3-3 所示。

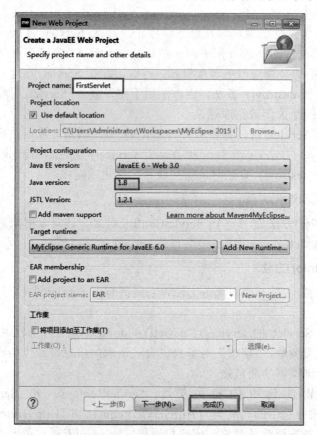

图 3-3 输入工程名称

单击图 3-3 中的"完成"按钮，将返回 MyEclipse 的主界面，在主界面的左边将出现工程项目的结构。

第二步：新建 Servlet。在 FirstServlet 工程的 src 结点上右击，选择"新建"→Servlet 命

第 3 章　Servlet 技术详解

令，如图 3-4 所示。在弹出的 Create a new Servlet 对话框中进行配置，如图 3-5 和图 3-6 所示。

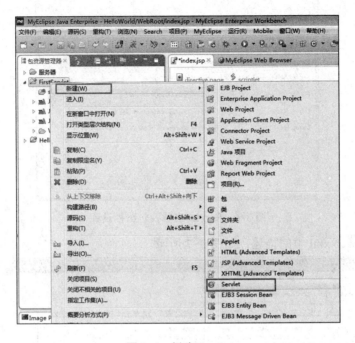

图 3-4　新建 Servlet

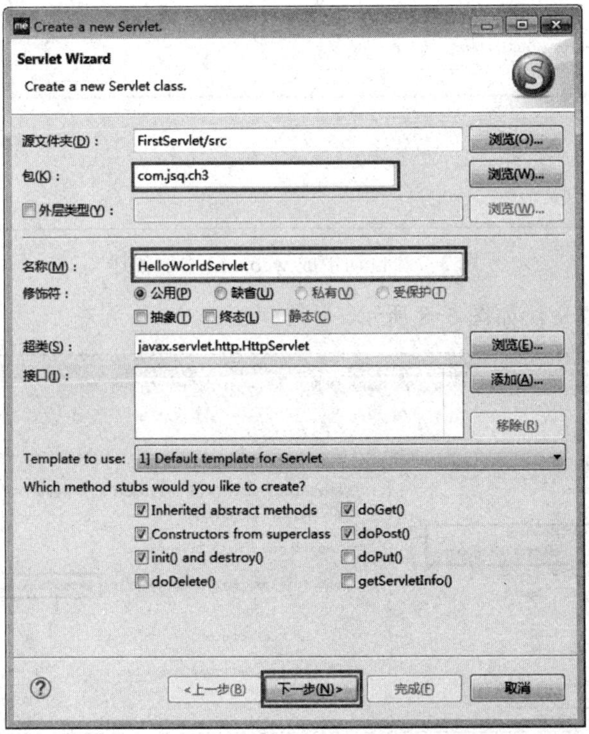

图 3-5　工程目录结构

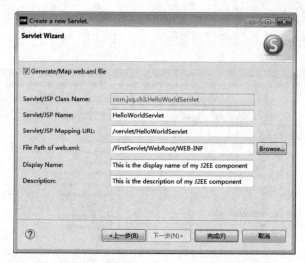

图 3-6　Servlet 名称及包路径设置

自动生成 web.xml 中的描述，如图 3-7 所示。

图 3-7　自动生成 web.xml 中的描述

工程目录发生改变，如图 3-8 所示。

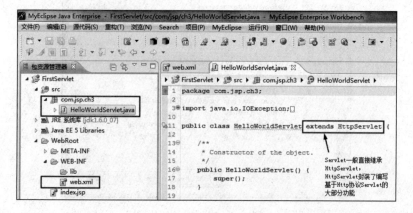

图 3-8　工程目录发生改变

HelloWorldServlet.java 文件代码如下：

```java
package com.jsq.ch3;

import java.io.IOException;
import java.io.PrintWriter;

import javax.servlet.ServletException;
import javax.servlet.http.HttpServlet;
import javax.servlet.http.HttpServletRequest;
import javax.servlet.http.HttpServletResponse;

public class HelloWorldServlet extends HttpServlet {

    /**
     * Constructor of the object.
     */
    public HelloWorldServlet() {
        super();
    }

    /**
     * Destruction of the servlet. <br>
     */
    public void destroy() {
        super.destroy(); // Just puts "destroy" string in log
        //Put your code here
    }

    /**
     * The doGet method of the servlet. <br>
     *
     * This method is called when a form has its tag value method equals to get.
     *
     * @param request the request send by the client to the server
     * @param response the response send by the server to the client
     * @throws ServletException if an error occurred
     * @throws IOException if an error occurred
     */
    public void doGet(HttpServletRequest request, HttpServletResponse response)
            throws ServletException, IOException {

        response.setContentType("text/html");
        PrintWriter out = response.getWriter();
        out.println("<!DOCTYPE HTML PUBLIC \"-//W3C//DTD HTML 4.01 Transitional//EN\">");
        out.println("<HTML>");
        out.println("  <HEAD><TITLE>A Servlet</TITLE></HEAD>");
```

```
            out.println("  <BODY>");
            out.print("    This is ");
            out.print(this.getClass());
            out.println(", using the GET method");
            out.println("  </BODY>");
            out.println("</HTML>");
            out.flush();
            out.close();
        }

        /**
         * The doPost method of the servlet. <br>
         *
         * This method is called when a form has its tag value method equals to post.
         *
         * @param request the request send by the client to the server
         * @param response the response send by the server to the client
         * @throws ServletException if an error occurred
         * @throws IOException if an error occurred
         */
        public void doPost(HttpServletRequest request, HttpServletResponse response)
                throws ServletException, IOException {

            response.setContentType("text/html");
            PrintWriter out = response.getWriter();
            out.println("<!DOCTYPE HTML PUBLIC \"-//W3C//DTD HTML 4.01 Transitional//EN\">");
            out.println("<HTML>");
            out.println("  <HEAD><TITLE>A Servlet</TITLE></HEAD>");
            out.println("  <BODY>");
            out.print("    This is ");
            out.print(this.getClass());
            out.println(", using the POST method");
            out.println("  </BODY>");
            out.println("</HTML>");
            out.flush();
            out.close();
        }
        /**
         * Initialization of the servlet. <br>
         *
         * @throws ServletException if an error occurs
         */
        public void init() throws ServletException {
            //Put your code here
        }

    }
```

为了简化开发，缩写的 Servlet 一般直接继承 HttpServlet，HttpServlet 封装了编写基于 HTTP 协议 Servlet 的大部分功能。HelloWorldServlet 中有两个方法：doGet()和 doPost()，都进行一样的处理。在 doGet()方法中，首先设置响应的 MIME 类型和编码方式，然后获得输出流对象，这个输出流对象用 out 表示，这个 out 对象和 JSP 中内建对象 out 是一样的，通过它可以输出发送到客户端。

下面看一下在 WebRoot\WEB-INF\web.xml 文件中如何描述这个 Servlet。

web.xml 文件代码：

```xml
<?xml version="1.0" encoding="UTF-8"?>
<web-app version="3.0"
    xmlns="http://java.sun.com/xml/ns/javaee"
    xmlns:xsi="http://www.w3.org/2001/XMLSchema-instance"
    xsi:schemaLocation="http://java.sun.com/xml/ns/javaee http://java.sun.com/xml/ns/javaee/web-app_3_0.xsd">
    <!-- =======================开始=========================== -->
  <servlet>
    <description>This is the description of my J2EE component</description>
    <display-name>This is the display name of my J2EE component</display-name>
    <servlet-name>HelloWorldServlet</servlet-name>
    <servlet-class>com.jsq.ch3.HelloWorldServlet</servlet-class>
  </servlet>

  <servlet-mapping>
    <servlet-name>HelloWorldServlet</servlet-name>
    <url-pattern>/servlet/HelloWorldServlet</url-pattern>
  </servlet-mapping>
    <!-- =======================结束=========================== -->
</web-app>
```

第三步：部署工程文件，如图 3-9～图 3-12 所示。

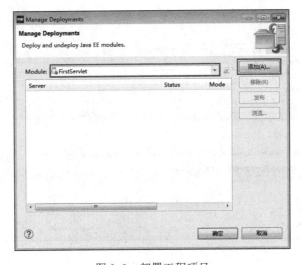

图 3-9　部署工程项目

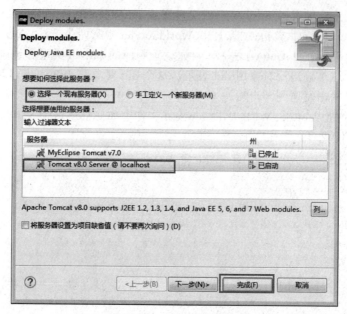

图 3-10　配置部署模块

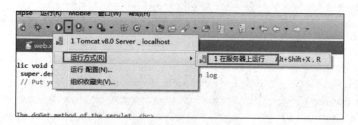

图 3-11　选择"在服务器上运行"命令

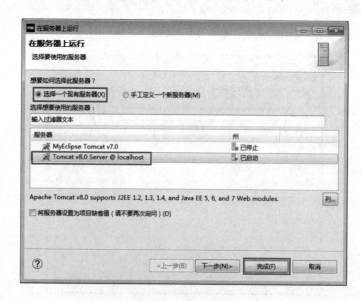

图 3-12　选择 Web 容器

第 3 章 Servlet 技术详解

第四步：在浏览器中预览结果，如图 3-13 所示。

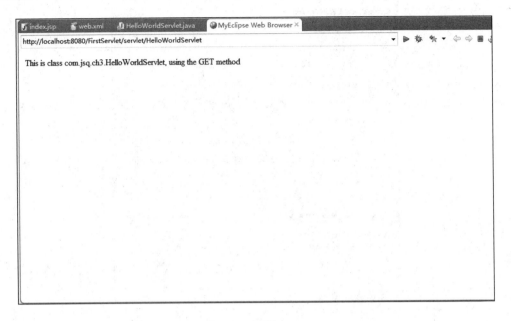

图 3-13 运行 Servlet

3.2 站点计数监听器制作

在许多触发性的处理中需要监听功能。通常用作用户某一事件的触发监听，比如监听用户的来访与退出、监听某一数据事件的发生，或者定义一个周期性的时针定期执行。这一功能极大地增强了 Java Web 程序的事件处理能力。

下面开发一个在线用户计数器。

第一步：启动 MyEclipse，新建一个 Web 工程，如图 3-14 和图 3-15 所示。

图 3-14 新建 Web 工程

-47-

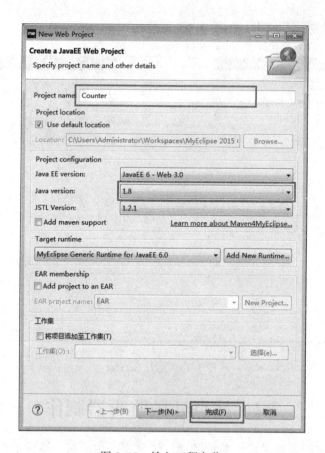

图 3-15 输入工程名称

第二步：新建一个监听类 CounterListener.java，如图 3-16～图 3-18 所示。

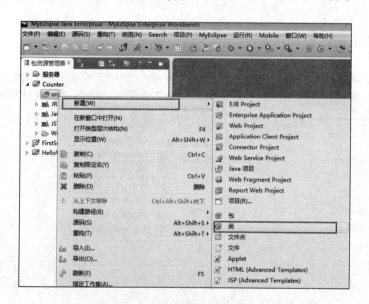

图 3-16 新建类文件

第 3 章 Servlet 技术详解

图 3-17 命名类　　　　　　　　　　图 3-18 工程结构

CounterListener.java 监听类的代码如下：

```java
package com.utils;
import javax.servlet.http.HttpSessionEvent;
import javax.servlet.http.HttpSessionListener;
/*注意：该类继承 HttpSessionListener */
public class CounterListener implements HttpSessionListener {
    public static int count;
    public CounterListener()
    {
        count=0;
    }
    public void sessionCreated(HttpSessionEvent se)
    {
        count++;
    }
    public void sessionDestroyed(HttpSessionEvent se)
    {
        if(count>0)
            count--;
    }
}
```

在上面的代码中，变量 count 为静态变量，在整个系统中唯一，记录整个系统中在线

-49-

用户数，sessionCreated()在用户到访时自动调用，使得统计数加 1；sessionDestroyed()在用户会话过期或单击退出销毁 session 时调用，使得统计数减 1。

第三步：打开 WebRoot/WEB-INF/web.xml 文件，添加如下的监听器配置代码：

```xml
<?xml version="1.0" encoding="UTF-8"?>
<web-app version="2.5"
    xmlns="http://java.sun.com/xml/ns/javaee"
    xmlns:xsi="http://www.w3.org/2001/XMLSchema-instance"
    xsi:schemaLocation="http://java.sun.com/xml/ns/javaee
    http://java.sun.com/xml/ns/javaee/web-app_2_5.xsd">
  <welcome-file-list>
    <welcome-file>index.jsp</welcome-file>
  </welcome-file-list>
<!--添加监听器listener-->
  <listener>
    <listener-class>com.utils.CounterListener</listener-class>
  </listener>
</web-app>
```

第四步：在 index.jsp 中添加一条语句，用于显示当前在线用户数：

```
当前在线用户数为:<%=CounterListener.count %>
```

第五步：布置项目，并启动 Tomcat，在浏览器地址栏中输入 http://localhost:8080/counter/index.jsp，即可预览效果。

小　　结

本章主要围绕 Servlet 技术介绍了相关的概念和开发实例。

Servlet 指服务器端小程序，它是一种很成熟的技术，它先于 Java EE 平台出现。从本质上讲，Servlet 就是一个 Java 类，Java 语言能够实现的功能，Servlet 基本上都能实现（除图形界面外）。Servlet 主要用于处理客户机传来的 HTTP 请求，并返回一个响应。通常所讲的 Servlet 就是指 HttpServlet。在开发 Servlet 时，可以直接继承 javax.servlet.http.HttpServlet。Servlet 需要在 web.xml 中进行描述（如果使用 MyEclipse 开发工具创建，此描述会自动添加），描述一般包括 Servlet 名字、Servlet 类、初始参数、安全配置、URI 映射等。Servlet 可以生成 HTML 脚本输出。

习　　题

新建一个工程，工程名称为 Ex3，使用 JSP+Servlet 组合完成第 2 章的用户注册功能（提示：将信息接收页面改为使用 Servlet 实现）。

第 4 章
JSP 中使用 JavaBean

学习目标
- 了解什么是 JavaBean。
- 掌握 JavaBean 的开发与使用。
- 了解 JavaDoc 文档的生成。

本章主要讲述了 JavaBean 在 JSP 中应用的相关问题。首先讲述了 JavaBean 的基本概念，接着讲述了 JavaBean 的属性和方法，介绍了 JavaBean 的开发。

4.1 JavaBean 简介

JSP 最强有力的一个方面就是能够使用 JavaBean 组件体系。Javabean 往往封装了程序的页面逻辑，它是可重用的组件。通过使用 JavaBean，可以减少在 JSP 中脚本代码的使用，这样使用 JSP 更加易于维护，易于被非编程人员接受。

JavaBean 体系结构是第一个全面基于组件的标准模型之一。JavaBean 是描述 Java 的软组件模型，有点类似于 Microsoft 的 com 组件。JavaBean 组件是 Java 类，这些类遵循一个接口格式，以便于使方法命名、底层行为以及继承或实现的行为能够把类看作标准的 JavaBean 组件进行构造。

JavaBean 具有以下特性：
（1）可以实现代码的重复利用；
（2）易维护性、易使用性、易编写性；
（3）可以在支持 Java 的任何平台上工作，而不需要重新编译；
（4）可以在内部、网内或者是网络之间进行传输；
（5）可以以其他部件的模式进行工作。

JavaBean 分为可视化和非可视化。如 AWT 下的应用就是可视化领域。现在，JavaBean 更多地应用于非可视化领域，它在服务端应用方面表现出了越来越强的生命力。非可视化 JavaBean 在 JSP 程序中常用来封装事务逻辑、数据库操作等，可以很好地实现业务逻辑和前台程序的分离，使得系统具有更好的健壮性和灵活性。

4.1.1　JavaBean 的属性

JavaBean 的属性与一般 Java 程序中所指的属性，或者说与所有面向对象的程序设计语言中对象的属性是一个概念，在程序中的具体体现就是类中的变量。在 JavaBean 设计中，按照属性的不同作用又细分为四类：Simple（简单的）、Index（索引的）、Bound（绑定的）与 Constrained（约束的）。

1. Simple（简单的）

一个简单属性表示一个伴随有一对 get()/set()方法（C 语言的过程或函数在 Java 程序中称为"方法"）的变量。属性名与和该属性相关的 get/set 方法名对应。例如：如果有 setX() 和 getX()方法，则暗指有一个名为"X"的属性。如果有一个方法名为 isX()，则通常暗指"X"是一个布尔属性（即 X 的值为 true 或 false）。

简单属性的 JavaBean 代码：

```
package com..ch4.javabean;
public class SimpleBean{
private String  name= "Jim";       //属性名为 name，类型为字符串
private boolean  active=false;     //属性名为 active，类型为布尔类型
/* getXXX()方法，返回这个属性的值*/
public String getName()
{
   return this.name;
}
/* setXXX()方法，设置这个属性的值*/
public void setName(String name)
{
   this.name=name;
}
//对于 Boolean 类型的属性，可以使用 isXXX()方法来获得属性
public boolean  isActive()
{
   return this .active;
}
//设置 Boolean 类型的属性
public void setActive(boolean active)
{
   this.active=active;
}
}
```

2. Indexed（索引的）

一个 Indexed 属性表示一个数组值。使用与该属性对应的 set()/get()方法可取得数组中的数值。该属性也可一次设置或取得整个数组的值。

索引属性的 JavaBean 代码：

```
package com..ch4.javabean;
public class IndexedBean{
    int[] dataSet{1,2,3,4,5,6};
```

第 4 章 JSP 中使用 JavaBean

```
    public void setDataSet(int[] x){
        dataSet = x;
    }
    public void setDataSet(int index,int x){
        dataSet[index] = x;
    }
    public int[] getDataSet(){
        return dataSet;
    }
}
```

有了以上的 JavaBean，便可通过以下方式使用该 JavaBean：

```
setDataSet(4,8);
int dataSet = getDataSet(5);
int[] s= getDataSet();
```

3．Bound（绑定的）

一个 Bound 属性是指当该种属性的值发生变化时，要通知其他的对象。每次属性值改变时，这种属性就触发一个 PropertyChange 事件（在 Java 程序中，事件也是一个对象）。事件中封装了属性名、属性的原值、属性变化后的新值。这种事件是传递到其他的 JavaBean，至于接收事件的 JavaBean 应做什么动作由其自己定义。

此属性在 JavaBean 图形编程中大量使用，本书重点讲述在 JSP 中如何使用 JavaBean，故不作详细介绍。

4．Constrained（约束的）

一个 JavaBean 的 constrained 属性，是指当这个属性的值要发生变化时，与这个属性已建立了某种连接的其他 Java 对象可否决属性值的改变。constrained 属性的监听者通过抛出 PropertyVetoException 来阻止该属性值的改变。

此处只介绍该属性的基本概念，在 JSP 开发中很少使用，故不作详细介绍。

4.1.2 JavaBean 的方法

JavaBean 的方法的编写和其他 Java 程序的方法一样。例如：

```
public void showMessage(){
    System.out.println("Message title:"+title);
    System.out.println("Message content:"+ content);
}
```

上面的代码就是一个 JavaBean 方法，它只是标准的 Java 代码。

4.2 创建一个 JavaBean

下面通过一个例子设计一个简单的计算器，要求实现功能：加、减、乘、除运算。

第一步：在工程文件 src 上新建一个类，命名为 Calculator.java，如图 4-1 所示。

基础篇

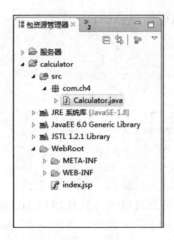

图 4-1　工程结构图

第二步：在 Calculator.java 类中进行属性声明，代码如下所示。

```
package com.ch4;

public class Calculator {
    private String value1;          /*表单中的第一个参数*/
    private String value2;          /*表单中的第二个参数*/
    private double result;          /*计算结果*/
    private String operatorChar;    /*运算符号*/
}
```

第三步：利用 MyEclipse 自动生成相应的 Getter()与 Setter()方法，如图 4-2 和图 4-3 所示。

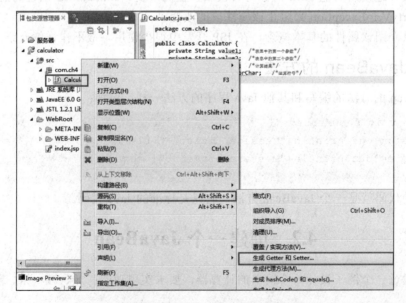

图 4-2　自动生成 Getter()和 Setter()方法

第 4 章　JSP 中使用 JavaBean

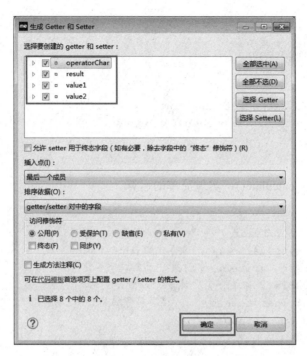

图 4-3　开始自动生成 Getter 和 Setter 方法

单击图 4-3 中的"确定"按钮后，MyEclipse 中的代码将发生如下改变。

```java
package com.ch4;
public class Calculator {
    private String value1="0";          /*表单中的第一个参数*/
    private String value2="0";          /*表单中的第二个参数*/
    private double result=0;            /*计算结果*/
    private String operatorChar="";     /*运算符号*/
    /*下面代码均为自动生成的*/
    public String getValue1() {
        return value1;
    }
    public void setValue1(String value1) {
        this.value1=value1;
    }
    public String getValue2() {
        return value2;
    }
    public void setValue2(String value2) {
        this.value2=value2;
    }
    public double getResult() {
        return result;
    }
    public void setResult(double result) {
```

```java
        this.result = result;
    }
    public int getOperatorNum() {
        return operatorNum;
    }
    public void setOperatorNum(int operatorNum) {
        this.operatorNum = operatorNum;
    }
}
```

第四步：自定义方法 calculate()，实现简单的运算。

```java
package com.ch4;
public class Calculator {
    private String value1;    /*表单中的第一个参数*/
    private String value2;    /*表单中的第二个参数*/
    private double result;    /*计算结果*/
    private int operatorNum;  /*运算符号编号*/
    public String getValue1() {
        return value1;
    }
    public void setValue1(String value1) {
        this.value1=value1;
    }
    public String getValue2() {
        return value2;
    }
    public void setValue2(String value2) {
        this.value2=value2;
    }
    public double getResult() {
        return result;
    }
    public void setResult(double result) {
        this.result=result;
    }
    public int getOperatorNum() {
        return operatorNum;
    }
    public void setOperatorNum(int operatorNum) {
        this.operatorNum=operatorNum;
    }
    /**
     * 根据不同的操作符号编码，进行不同的运算
     **/
    public void calculate(){
```

```
    double a=Double.parseDouble(value1);
    double b=Double.parseDouble(value2);
    try {
        int key=operatorNum;/*注意,此处operatorNum是int型,在设置表单时,
                                                     提交的也应当是int型*/
        switch (key) {
        case 1:
            result=a+b;
            break;
        case 2:
            result=a-b;
            break;
        case 3:
            result=a*b;
            break;
        case 4:
            result=a/b;
            break;
        default:
            break;
        }
    } catch (Exception e) {
        System.out.println(e);
    }
}
```

在上面的代码中，value1 与 value2 分别表示两个操作数，操作符号的标识用 operatorNum 表示。calculate()方法用于计算，并将计算后的结果保存在 result 属性中，最后通过 getResult()方法获得计算的结果。

至此，一个简单的计算器 JavaBean 创建完毕。在 4.3 节中，将通过 JSP 文件进行此 JavaBean 的调用。

4.3 在 JSP 中调用 JavaBean

第一步：使用 Dreamweaver（界面布局建议使用此软件，有利于提高效率）打开 4.2 节中的工程文件 index.jsp。创建如图 4-4 和图 4-5 所示的操作界面。

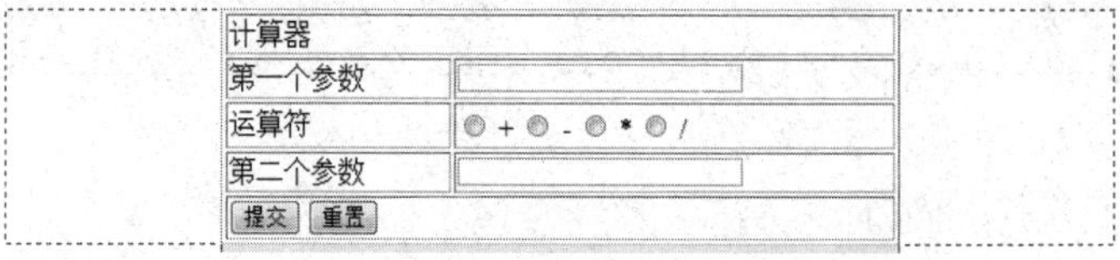

图 4-4 index.jsp 页面

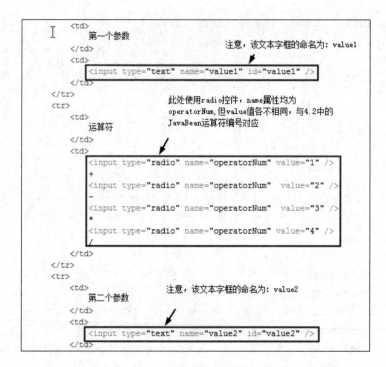

图 4-5 index.jsp 页面标签特别说明

第二步：通过<jsp:useBean></jsp:useBean>标签使用 JavaBean，具体代码如下：

```
<jsp:useBean id="cal" scope="request" class="com.ch4.Calculator">
</jsp:useBean>
```

其中 id="cal"好比给类 Calculator 取了一个别名，以后要调用 Calculator 中的方法就通过 id 中的值"cal"来调用。scope="request"指定 Bean 的范围为 request。

第三步：<jsp:setProperty>用来设置已经实例化的 Bean 对象的属性，具体代码如下：

```
<jsp:setProperty property="*" name="cal"/>
```

该形式是设置 Bean 属性的快捷方式。在 Bean 中属性的名字，类型必须和 request 对象中的参数名称相匹配。由于表单中传过来的数据类型都是 String 类型的，JSP 内在机制会把这些参数转化成 Bean 属性对应的类型。property ="*"表示所有名字和 Bean 属性名字匹配的请求参数都将被传递给相应的属性 set()方法。

```
<%@ page language="java" import="java.util.*" pageEncoding="utf-8"%>
    <jsp:useBean id="cal" scope="request"
class="com.ch4.Calculator"></jsp:useBean>
    <jsp:setProperty property="*" name="cal"/>
    <!DOCTYPE HTML PUBLIC "-//W3C//DTD HTML 4.01 Transitional//EN">
    <html>
        <head>
            <title>Jsp 中使用 JavaBean</title>
        </head>
        <body>
```

第 4 章 JSP 中使用 JavaBean

```html
<form id="form1" name="form1" method="post" action="index.jsp">
    <table width="399" border="1">
        <tr>
            <td colspan="2">
                计算器
            </td>
        </tr>
        <tr>
            <td>
                第一个参数
            </td>
            <td>
                <input type="text" name="value1" id="value1" />
            </td>
        </tr>
        <tr>
            <td>
                运算符
            </td>
            <td>
                <input type="radio" name="operatorNum" value="1" />+
                <input type="radio" name="operatorNum" value="2" />-
                <input type="radio" name="operatorNum" value="3" />*
                <input type="radio" name="operatorNum" value="4" />/
            </td>
        </tr>
        <tr>
            <td>
                第二个参数
            </td>
            <td>
                <input type="text" name="value2" id="value2" />
            </td>
        </tr>
        <tr>
            <td colspan="2">
<input type="submit" name="button" id="button" value="提交" />
<input type="reset" name="button2" id="button2" value="重置" />
            </td>
        </tr>
    </table>
</form>
<%
cal.calculate();
out.println("运算结果: "+cal.getResult());

%>
</body>
</html>
```

-59-

第四步：先部署工程，并启动 Tomcat，如图 4-6 和图 4-7 所示。

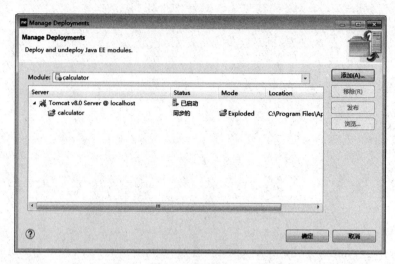

图 4-6　部署工程

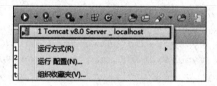

图 4-7　启动 Tomcat 8

第五步：在浏览器中输入 http://localhost:8080/calculator/index.jsp，即可进行计算，测试结果如图 4-8 所示。

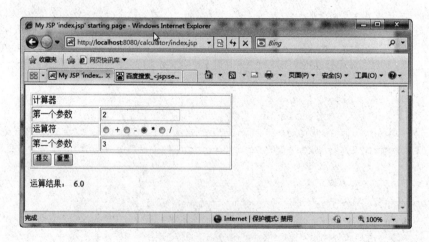

图 4-8　计算结果测试

4.4　JavaDoc 文档的生成

JavaDoc 是 Sun 公司提供的一个技术，它从程序源代码中抽取类、方法、成员等注释

第 4 章 JSP 中使用 JavaBean

形成一个和源代码配套的 API 帮助文档。也就是说，只要在编写程序时以一套特定的标签作注释，在程序编写完成后，通过 JavaDoc 就可以同时形式程序的开发文档了。JavaDoc 输出的是一些静态网页文档，经过 Web 浏览器来查看。

/** …… */用于注释若干行，并写入 JavaDoc 文档，如图 4-9～图 4-15 所示。

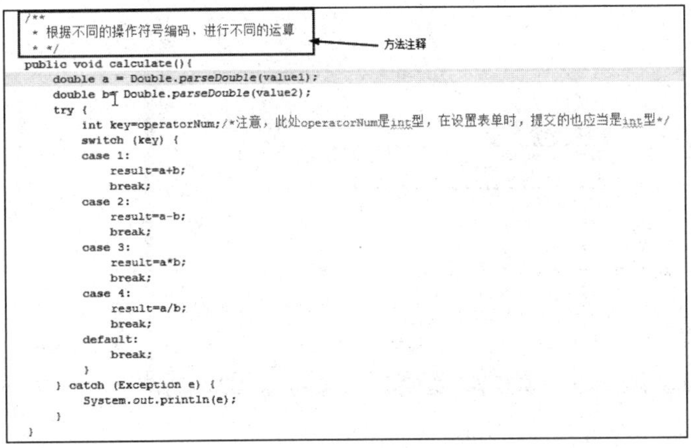

图 4-9　注释代码

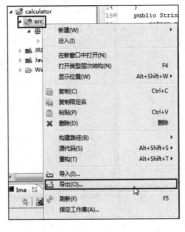

图 4-10　导出数据

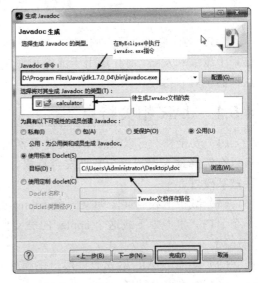

图 4-11　选择导出类型为 Javadoc

图 4-12　导出 Javadoc 配置

图 4-13　导出 Javadoc 日志记录

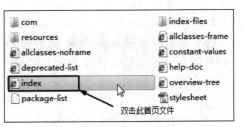

图 4-14　浏览 Javadoc 文档

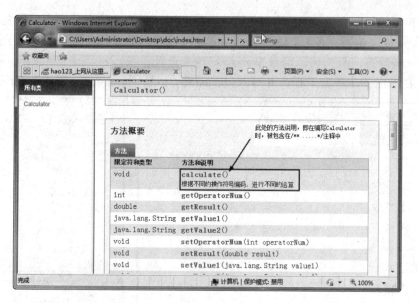

图 4-15 查看生成的 Javadoc 文档注释

4.5 JAR 插件的制作与使用

JAR 文件格式以流行的 ZIP 文件格式为基础。与 ZIP 文件不同的是，JAR 文件不仅用于压缩和发布，而且还用于部署和封装库、组件和插件程序，并可被像编译器和 JVM 这样的工具直接使用。

4.5.1 JAR 相关特点

（1）安全性。可以对 JAR 文件内容加上数字化签名。这样，能够识别签名的工具就可以有选择地为用户授予软件安全特权，这是其他文件做不到的，它还可以检测代码是否被篡改过。

（2）减少下载时间。如果一个 applet 捆绑到一个 JAR 文件中，那么浏览器就可以在一个 HTTP 事务中下载这个 applet 的类文件和相关的资源，而不是对每一个文件打开一个新连接。

（3）压缩。JAR 格式允许用户压缩文件以提高存储效率。

（4）传输平台扩展。Java 扩展框架（Java Extensions Framework）提供了向 Java 核心平台添加功能的方法，这些扩展是用 JAR 文件打包的（Java 3D 和 JavaMail 就是由 Sun 开发的扩展例子）。

（5）包密封。存储在 JAR 文件中的包可以选择进行密封，以增强版本一致性和安全性。密封一个包意味着包中的所有类都必须在同一 JAR 文件中找到。

（6）包版本控制。一个 JAR 文件可以包含有关它所包含的文件的数据，如厂商和版本信息。

（7）可移植性。处理 JAR 文件的机制是 Java 平台核心 API 的标准部分。（见图 4-16～图 4-18）

第 4 章　JSP 中使用 JavaBean

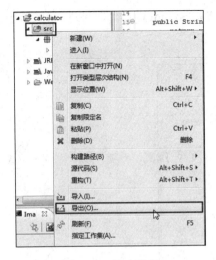

图 4-16　导出数据

图 4-17　导出 JAR 文件

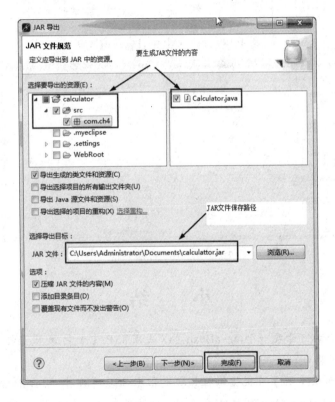

图 4-18　JAR 文件保存路径

4.5.2　JAR 的使用

第一步：新建一个 Web Project 工程，在该工程的 WebRoot/WEB-INF/lib 目录下粘贴 4.5.1 节中生成的 calculator.jar 插件，如图 4-19 所示。

第二步：将 4.3 节中的 index.jsp 复制到该项目的 WebRoot 目录覆盖，并部署启动运行

Tomcat，在浏览器中输入 http://localhost:8080/test/index.jsp，即可进行计算，测试结果如图 4-20 所示。

图 4-19　导入 JAR 插件到新项目中

图 4-20　计算结果测试

小　　结

本章主要围绕如何开发一个 JavaBean，以及在 JSP 中如何使用 JavaBean，介绍了相关的概念和开发实例。

JavaBean 组件主要用于可视化编程领域，但在 JSP 开发中，JavaBean 主要用于支持后台业务逻辑处理。JavaBean 往往封装了 JSP 的业务逻辑，它是可重用的组件。只用 JSP 开发会使得 JSP 文件非常混乱；如果使用了 JavaBean，则可以大大减少 JSP 中代码量。JSP 的理想状态是只负责显示，而不负责处理。在 4.4 节与 4.5 节中主要讲了 Javadoc 文档与 JAR 插件的制作，在许多企业的团队开发中，项目组长会分发项目组成员一份本系统的 JAR 插件，以及针对该插件的使用说明文档，即 Javadoc 文档。另外，JavaBean 最为常用的是连接与操作数据库。在第 6 章有详细介绍。

习 题

1. 新建一个工程 Ex4，在此工程中创建一个 JavaBean，用于进行字符串的加密（加密算法可自行设计，也可上网查询相关加密算法，如 MD5 等）。

2. 在 Ex4 中的 index.jsp 文件中设计一个表单，如图 4-21 所示，单击"加密"按钮后，将原字符串通过加密类进行加密并输出显示。

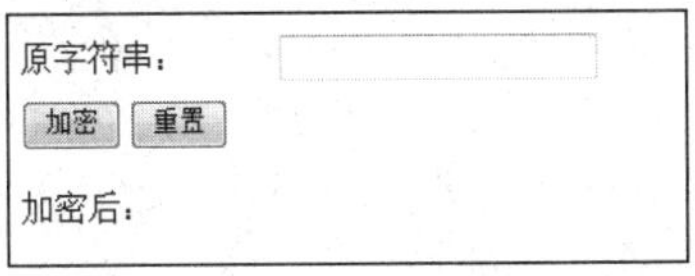

图 4-21 表单

补充：MD5 加密码类文件。

```java
package com.common;

import java.io.UnsupportedEncodingException;
import java.security.MessageDigest;
import java.security.NoSuchAlgorithmException;

/**
 * MD5 加密类
 */
public class MD5 {

    private static MessageDigest digest=null;

    public synchronized static final String Encrypt(String data){
        return Encrypt(data, 16);
    }
    public synchronized static final String Encrypt(String data, int len){
        if(digest==null) {
            try {
                digest = MessageDigest.getInstance("MD5");
            }
            catch (NoSuchAlgorithmException e) {
                e.printStackTrace();
            }
        }
        if(len!=16 && len!=32) len=16;
        try {
            digest.update(data.getBytes("UTF-8"));
        } catch (UnsupportedEncodingException e) {
        }
        String s=encodeHex(digest.digest());
        if(len==16){
            return s.substring(8, 24);
```

```java
        }
        return s;
    }

    private static final String encodeHex(byte[] bytes) {
        int i;
        StringBuffer buf=new StringBuffer(bytes.length*2);
        for(i=0; i < bytes.length; i++) {
            if(((int) bytes[i] & 0xff) < 0x10) {
                buf.append("0");
            }
            buf.append(Long.toString((int) bytes[i] & 0xff, 16));
        }
        return buf.toString();
    }
}
```

第 5 章 搭建数据库开发环境

学习目标
- 安装 MySQL 数据库。
- 安装 Navicat for MySQL 客户端软件。
- 创建数据库、数据表、。
- 掌握常用的 SQL 语句。

本章介绍了如何安装及配置 MySQL 数据库，使用 Navicat 来创建数据库和数据表，并添加了样例数据。通过本章的学习，可以提高操纵 MySQL 数据库的熟练程度。

5.1 MySQL 概述

5.1.1 MySQL 简介

MySQL 是一个小型关系型数据库管理系统，开发者为瑞典 MySQL AB 公司。在 2008 年被 Sun 公司收购。而 2009 年，SUN 又被 Oracle 收购。MySQL 是一种关联数据库管理系统，关联数据库将数据保存在不同的表中，而不是将所有数据放在一个大仓库内。这样就增加了速度并提高了灵活性。MySQL 的 SQL 是指"结构化查询语言"，是用于访问数据库的最常用标准化语言。MySQL 软件采用了 GPL（GNU 通用公共许可证）。由于其体积小、速度快、总体拥有成本低，尤其是开放源码这一特点，因此许多中小型网站为了降低网站总体拥有成本而选择了 MySQL 作为网站数据库。

MySQL 的特性如下：

（1）使用 C 和 C++编写，并使用了多种编译器进行测试，保证源代码的可移植性。

（2）支持 AIX、FreeBSD、HP-UX、Linux、Mac OS、NovellNetware、OpenBSD、OS/2 Wrap、Solaris、Windows 等多种操作系统。

（3）为多种编程语言提供了 API。这些编程语言包括 C、C++、Python、Java、Perl、PHP、Eiffel、Ruby 和 Tcl 等。

（4）支持多线程，充分利用 CPU 资源。

（5）优化的 SQL 查询算法，有效地提高查询速度。

（6）既能够作为一个单独的应用程序应用在客户端服务器网络环境中，也能够作为一个库而嵌入其他软件中提供多语言支持，常见的编码如中文的 GB 2312、BIG5，日文的 Shift_JIS 等都可以用作数据表名和数据列名。

（7）提供 TCP/IP、ODBC 和 JDBC 等多种数据库连接途径。

（8）提供用于管理、检查、优化数据库操作的管理工具。

（9）可以处理拥有上千万条记录的大型数据库。

（10）支持多种存储引擎。

5.1.2 下载并安装 MySQL

截至本书编写完毕，MySQL 的最新版本为 5.5。本书后面所用到的数据库均为 MySQL 5.5。

1．下载 MySQL 数据库

MySQL 数据库的下载地址：http://www.mysql.com/downloads/（或者百度搜索关键词"MySQL5.5 下载"）。下载后文件为 mysql-5.5.23-win32.zip 。

2．安装 MySQL 数据库

双击压缩包中的 mysql-5.5.23-win32.msi 文件，打开 MySQL 数据库安装界面，如图 5-1 所示。单击图 5-1 中的 Next 按钮，进入许可协议界面，如图 5-2 所示。选择接受许可协议复选框，单击 Next 按钮。

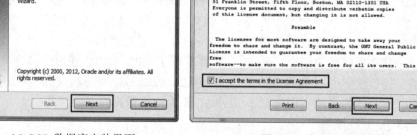

图 5-1　MySQL 数据库安装界面　　　　图 5-2　许可协议界面

为保证数据安全，防止因为重装系统而忘记备份数据库，造成数据丢失现象，建议不要将数据库文件保存在操作系统默认盘下（如：操作系统默认安装在 C 盘，那么 MySQL 数据库不要安装在 C 盘）。为此，将 MySQL 数据库的安装路径修改为 D:\ProgramData\MySQL\MySQL Server 5.5\，单击 Custom 按钮自定义安装（见图 5-3），进入图 5-4 所示的界面。设置安装路径，单击 Next 按钮，进入图 5-5 所示界面。单击 Install 按钮，进入图 5-6 所示界面。

第 5 章　搭建数据库开发环境

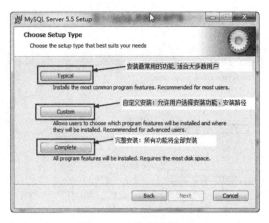

图 5-3　选择安装类型　　　　　图 5-4　自定义安装界面

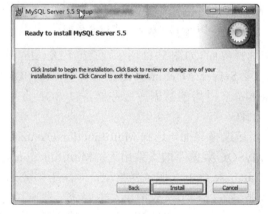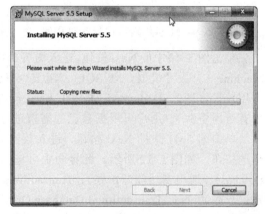

图 5-5　开始安装　　　　　　　图 5-6　安装进行中

单击图 5-7 中的 Next 按钮，进入图 5-8 所示界面，这里有一个配置向导的选项（Launch the MySQL Instance Configuration Wizard），建议勾选立即配置 MySQL，许多安装完 MySQL 后无法启动，原因就在于没有配置 MySQL，单击 Finish 按钮完成安装，并开始配置 MySQL，如图 5-8 所示。

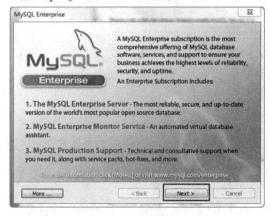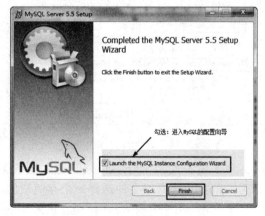

图 5-7　MySQL Enterprise 解说界面　　　图 5-8　完成安装

单击图 5-9 中的 Next 按钮，进入配置类型选择界面。选择 Detailed Configuration（详细配置）单选按钮，如图 5-10 所示。

图 5-9　MySQL 配置向导欢迎界面

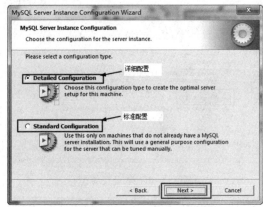

图 5-10　配置类型选择界面

单击图 5-10 中的 Next 按钮，进入服务类型选择界面，如图 5-11 所示，可选择 Developer Machine（开发者机器，MySQL 占用很少资源）、Server Machine（服务器类型，MySQL 占用较多资源）、Dedicated MySQL Server Machine（专用的数据库服务器，MySQL 占用所有可用资源）。用户可根据自己的需求进行选择。

单击图 5-11 中的 Next 按钮，进入数据库用途选择界面，选择 Multifunction Database 单选按钮，如图 5-12 所示。此界面中可选择 MySQL 数据库的大致用途，Multifunctional Database（通用多功能型，好）、Transactional Database Only（服务器类型，专注于事务处理，一般）、Non-Transactional Database Only（非事务处理型，较简单，主要做一些监控、记数用，对 MyISAM 数据类型的支持仅限于 non-transactional），用户可根据自己的用途进行选择。

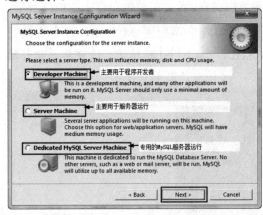

图 5-11　服务类型选择界面

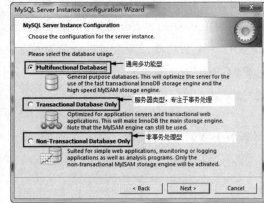

图 5-12　数据库用途选择界面

单击图 5-12 中的 Next 按钮，进入图 5-13 所示界面，对 InnoDB Tablespace 进行配置，即为 InnoDB 数据库文件选择一个存储空间，这里使用默认位置。

单击图 5-13 中的 Next 按钮，进入图 5-14 所示界面，选择 MySQL 的同时连接数。其中包括 Decision Support(DSS)/OLAP（20 个左右）、Online Transaction Processing(OLTP)

（500 个左右）、Manual Setting（手动设置，自己输一个数），这里选择 Online Transaction Processing(OLTP)，单击 Next 按钮。

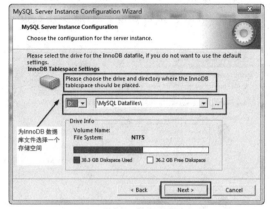

图 5-13　样式管理器界面

图 5-14　MySQL 连接数设置界面

单击图 5-14 中的 Next 按钮，进入图 5-15 所示界面，设置 MySQL，选择 Enable TCP/IP Networking 启用 TCP/IP 连接，设定端口 Port Number 为 3306。如果不启用 TCP/IP 连接，则只能在本机访问 MySQL 数据库。在本界面中，还可以选择 Enable Strict Mode（启用标准模式），这样 MySQL 就不会允许细小的语法错误。建议新手取消标准模式以减少麻烦。但熟悉 MySQL 以后，建议尽量使用标准模式，因为它可以降低有害数据进入数据库的可能性。

单击图 5-15 中的 Next 按钮，进入图 5-16 所示界面，设置 MySQL 数据库语言编码。建议使用多字节的通用 UTF8 编码，此编码对中英文存储等的支持较好。单击 Next 按钮，进入图 5-17 所示界面，勾选 Install As Windows Service 复选框，将 MySQL 安装为 Windows 服务，且指定 Service Name（服务标识名称）为默认值。勾选 Include Bin Directory in Windows PATH，将 MySQL 的 bin 目录加入 Windows PATH，加入后，就可以直接使用 bin 下的文件，而不用指出目录名，比如连接 mysql.exe -uusername –ppassword;便可，无须指出 mysql.exe 的完整地址，调用方便。

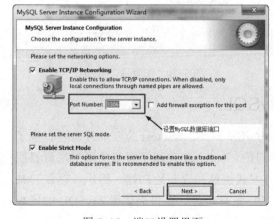

图 5-15　端口设置界面

图 5-16　编码设置界面

单击图 5-17 中的 Next 按钮，进入图 5-18 所示界面，勾选 Modify Security Settings

复选框，修改 MySQL 数据库的默认账号与密码，此处设置为 123456。Enable root access from remote machines 用于设置是否允许 root 用户在其他的机器上登录。Create An Anonymous Account 用于新建一个匿名用户，匿名用户可以连接数据库，不能操作数据，包括查询。

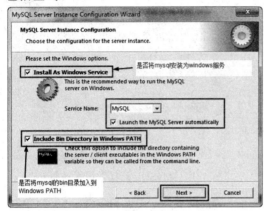
图 5-17 样式管理器界面

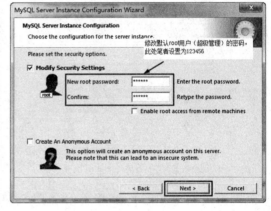
图 5-18 设置 MySQL 账户与密码界面

单击图 5-19 中的 Execute 按钮开始配置，当出现图 5-20 所示界面时，说明配置完成，单击 Finish 按钮即可，在 5.1.3 节将使用客户端软件 Navicat 软件连接 MySQL 服务器并操作其中的数据。

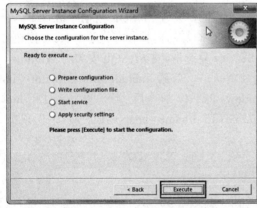
图 5-19 等待执行设置项界面

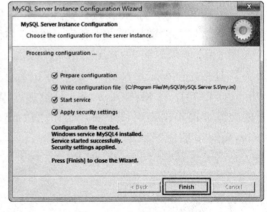
图 5-20 配置成功界面

5.1.3 下载并安装 Navicat for MySQL

Navicat 是一个强大的 MySQL 数据库管理和开发工具。Navicat 为专业开发者提供了一套强大的工具，且对于新用户仍然是易于学习的。Navicat 使用了极好的图形用户界面（GUI），可以让用户用一种安全和更为容易的方式快速和容易地创建、组织、存取和共享信息。Navicat 基于 Windows 平台，为 MySQL 量身定做，提供类似于 MySQL 的用户管理界面工具。此解决方案的出现，将解放 PHP、Java EE 等程序员以及数据库设计者、管理者的大脑，降低开发成本，带来更高的开发效率。用户可完全控制 MySQL 数据库

第 5 章 搭建数据库开发环境

和显示不同的管理资料，包括一个多功能的图形化管理用户和访问权限的管理工具，方便将数据从一个数据库移转到另一个数据库中（Local to Remote、Remote to Remote、Remote to Local），进行文档备份。Navicat 支持 Unicode，以及本地或远程 MySQL 伺服器连线，用户可浏览数据库、建立和删除数据库、编辑数据、建立或执行 SQL queries、管理用户权限（安全设定）、将数据库备份/复原、导入/导出数据（支持 CSV、TXT、DBF、和 XML 文件类型）等。新版与任何 MySQL 5.0.x 伺服器版本兼容，支持 Triggers，以及 BINARY VARBINARY/BIT 数据。

本书使用是的 Navicat for MySQL（MySQL 数据库管理工具）V10.0.11.0 简体中文版。相对 MySQL 安装，安装 Navicat 的安装简单许多，在安装向导中单击"安装"按钮即可，如图 5-21 所示。

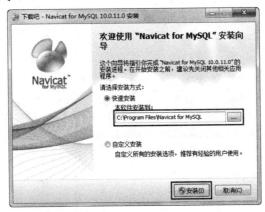

图 5-21　Navicat for MySQL 安装向导

5.2　使用 MySQL 数据库

5.2.1　采用 Navicat 管理 MySQL 数据库

（1）运行 Navicat for MySQL 管理软件，并创建一个连接，如图 5-22 和图 5-23 所示。

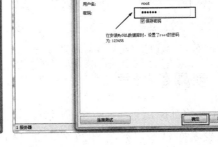

图 5-22　Navicat for MySQL 主界面　　　　图 5-23　创建连接

-73-

（2）单击图 5-23 中的"完成"按钮后，将返回 Navicat for MySQL 的主界面，双击左边的树形导航 localhost，出现如图 5-24 所示界面。

图 5-24　MySQL 自带数据库

5.2.2　创建数据库

在 localhost 连接上右击，选择"新建数据库"命令，输入数据库名称 shop 及使用的字符集编码 UTF-8 Unicode，如图 5-25 和图 5-26 所示。

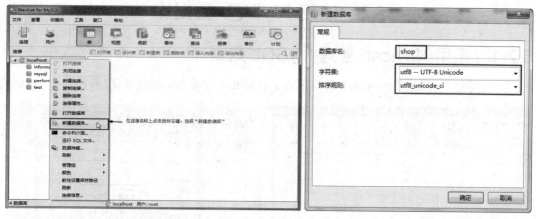

图 5-25　选择"新建数据库"命令　　　　图 5-26　设置新建数据库常规选项

5.2.3　创建数据表

在 shop 数据库结点下的"表"上右击，选择"新建表"命令，先创建字段，然后保存，输入表名称 User，如图 5-27 和图 5-28 所示。

第 5 章　搭建数据库开发环境

图 5-27　新建表

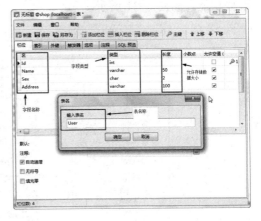

图 5-28　设计表结构

5.2.4　新增记录

双击 shop 数据库下的表 user 可以追加记录，如图 5-29 和图 5-30 所示。

图 5-29　查看表

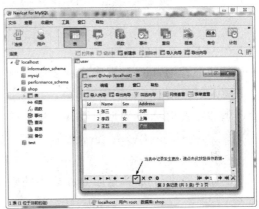

图 5-30　表中添加记录

5.3　SQL 语法介绍

5.3.1　SQL 简介

SQL（Structured Query Language，结构化查询语言）是一种数据库查询和程序设计语言，用于存取数据以及查询、更新和管理关系数据库系统。同时，.sql 也是数据库脚本文件的扩展名。

5.3.2　SQL 基本语法

1．数据查询

使用 Navicat for MySQL 中的 SQL Editor 的 SQL 执行功能，可以很方便地进行查询数据功能，如图 5-31 所示。

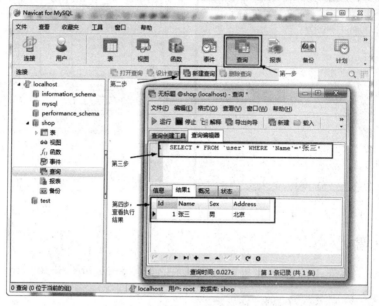

图 5-31 数据查询

（1）不带条件的查询。

例如：查询 shop 数据库 user 表中的所有记录。对应的 SQL 语句是：

```
select * from user
```

例如：查询 shop 数据库 user 表中的前 2 条记录。对应的 SQL 语句是：

```
select * from user limit 0,2
```

（2）带条件的查询。

例如：查询 shop 数据库 user 表中的姓名为"张三"的记录。对应的 SQL 语句是：

```
select * from user where name='张三'
```

例如：查询 shop 数据库 user 表中的性别是"男"的记录数。对应的 SQL 语句是：

```
select count(1) from user where sex='男'
```

例如：查询 shop 数据库 user 表中的姓"张"的学生记录。对应的 SQL 语句是：

```
select * from user where name like '张%'
```

2．数据添加

例如：在 shop 数据库 user 表中添加一个用户"赵六"，性别为"女"，地址为"桂林"。对应的 SQL 语句（见图 5-32）是：

```
Insert into user (name,sex,address) values ('赵六','女','桂林')
```

3．数据编辑

例如：将 shop 数据库 user 表中的用户"赵六"的性别改为"男"。对应的 SQL 语句（见图 5-33）是：

```
Update user set sex='男' where name='赵六'
```

第 5 章 搭建数据库开发环境

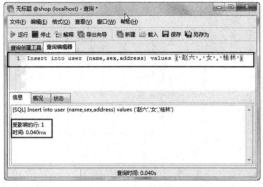

图 5-32 数据添加

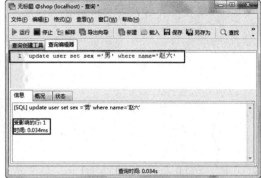

图 5-33 数据编辑

执行上面的 update 语句后，可以打开 user 表，检查数据是否修改成功。

4．数据删除

例如：将 shop 数据库 user 表中的用户"赵六"的性别改为"男"。对应的 SQL 语句（见图 5-34）是：

```
Delete from user where name='赵六'
```

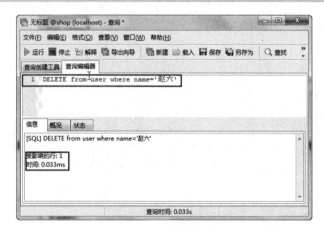

图 5-34 数据删除

执行上面的 delete 语句后，可以打开 user 表，检查数据是否修改成功。

小　　结

本章完成安装和配置了 MySQL 服务，并使用 Navicat for MySQL 来创建了数据库和数据表，并添加了样例数据。本章的创建过程也同时演示了在实际开发中该如何来方便地操纵 MySQL 数据库。

习　　题

1. 新建一个工程 Ex5，单击"学生注册"（见图 5-35）进入注册页面（见图 5-36），

-77-

并将数据保存到数据库中(建议密码进行加密处理,可参照第4章习题中的字符串加密)。

图 5-35 登录页面   图 5-36 注册页面

2. 实现登录验证,如果输入数据库中的用户名和密码正确则提示密码正确,否则提示用户名或密码不正确。

第 6 章　JDBC 技术详解

学习目标
- JDBC 技术和数据库驱动程序介绍。
- 使用 JDBC 连接各种数据库。
- JDBC 常用接口使用介绍。
- JDBC 操作数据库实例。

JSP 开发中离不开 JDBC 数据库编程，几乎所有的 JSP 项目均使用到数据库，所以掌握 JDBC 数据库编程技术非常重要。本章开始，将系统地介绍 JDBC 数据库编程技术。

6.1　JDBC 概述

JDBC 是一种用于执行 SQL 的 Java API，它是一组访问数据库的 API 集合，通过加载由数据库厂商所提供的驱动程序，可以与数据库建立连接。这样，就不必为访问不同的数据库而编写不同的程序了，只需用一个 JDBC API 编写一个程序就够了。而且 JDBC 技术对开发者屏蔽了一些细节问题，这样，程序员就不必去关心底层的实现技术。另外，与 Java 一样，JDBC 对数据库的访问也具有平台无关性。

JDBC 访问数据库的步骤：

第一步：注册并加载驱动程序。驱动程序是数据厂商所提供的一个对外的接口来分辨所访问的是哪一种数据库。我们可以调用 Class.forName()显式地加载驱动程序。方法如下（以 MySql 为例）：

```
Class.forName("org.gjt.mm.mysql.Driver");
```

Class.forName()是一个静态方法，用于指示 Java 虚拟机动态的查找、加载和链接指定类（如果尚未加载）。如果无法找到这个类，则抛出 ClassNotFountException。

第二步：创建连接。数据库连接是使用 DriverManager 对象的静态方法 getConnection()建立的，方法如下：

```
String url = "jdbc:mysql://127.0.0.1:3306/qq";    //127.0.0.1 为连接地
```

址,3306 为数据库端口,qq 为数据库名
```
    Connection conn = DriverManager.getConnection(url, "root", "123456");  //
"root", "root"为数据库用户名，123456 为数据库密码
```

第三步： 创建 SQL 语句对象。SQL 语句对象主要是用来执行 SQL 语句的，当建立了连接以后，便可以由 Statement 对象将 SQL 语句发送到 DBMS。对于 SELECT 语句，可以使用 executeQuery()；对于创建或修改表的语句，使用 executeUpdate()。方法如下：

```
    Statement st = conn.createStatement();
```

第四步： 提交 SQL 语句。得到创建 SQL 语句对象后，就可以提交 SQL 语句了，方法如下：

```
    ResultSet rs = st.executeQuery("select * from login");  //返回查询SQL语句
所得到的 ResultSet 对象格式的结果集
    int i = st.executeUpdate( "insert into login values('zhang','123')");   //
这个方法返回值是一个整数,代表的是影响数据的行数
```

第五步： 显示结果。当执行查询语句时，SQL 语句的执行结果存储在 ResultSet 对象中。可以使用 Getxxx()方法来检索数据。方法如下：

```
while(rs.next()){
    String name = set.getString("name");
    String password = set.getString("password");
}
```

如果发送的 SQL 语句不是查询语句，则这一步将省略。

第六步： 关闭连接。当完成数据库操作后，需要将连接关闭。因为数据库连接需要消耗系统资源。一旦不需要使用的时候，就应该释放出来。这是程序员必须养成的良好的习惯。在关闭连接时，正确的顺序是：Result、Statement、Connection。语法如下：

```
rs.close();
st.cloe();
conn.close();
```

为确保所有的连接能够被关闭，应该把关闭语句写在 finally 块中，这样，不管操作数据库过程中是否发生异常，都能够将资源释放出来。连接其他的数据库步骤均相同，只需更改驱动程序即可。

6.2　JDBC 数据库连接

6.2.1　连接 MySQL 数据库

第一步： 新建一个 Web Project，如图 6-1 所示。

第二步： 把 MySQL 的 JDBC 驱动程序复制到项目的 WEB-INF\lib 目录下，这个驱动的 JAR 包为  mysql-connector-java-5.1.6-bin.jar，如图 6-2 和图 6-3 所示。

第 6 章 JDBC 技术详解

图 6-1 新建 Web 工程

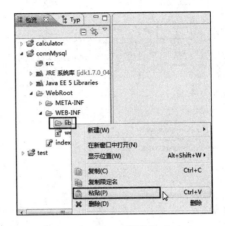

图 6-2 选择"粘贴"命令

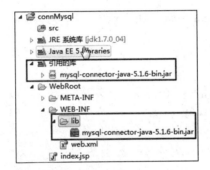

图 6-3 粘贴 mysql-connector-java-5.1.6-bin.jar 后工程目录结构

第三步：创建连接到 MySQL 数据库的 Java 类。在工程中的 src 目录上右击，选择"新建"→"类"命令，如图 6-4 所示。设置 Java 类，如图 6-5 所示。新建 Java 类后的工程结构图 6-6 所示。

图 6-4 新建 Java 类

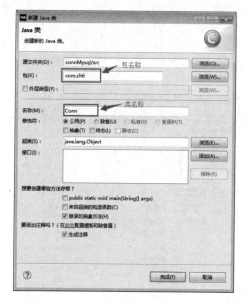

图 6-5 设置 Java 类

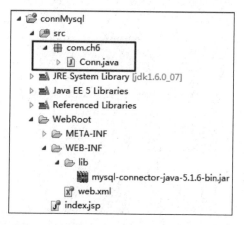

图 6-6 新建连接数据库类后的工程结构

第四步：在 com.ch6.Conn.java 中输入如下代码，实现连接 MySQL 数据库。代码如下：

```java
package com.ch6;
import java.sql.*;
public class Conn {
    Connection conn=null;
    Statement stmt=null;
    ResultSet rs=null;
    /**
     * 加载驱动程序
     **/
    public Conn() {
        try {
            Class.forName("com.mysql.jdbc.Driver");
        } catch (java.lang.ClassNotFoundException e) {
            System.err.println(e.getMessage());
        }
    }
    /**
     * 执行查询操作: select
     **/
    public ResultSet executeQuery(String sql) {
        try {
            conn = DriverManager
                    .getConnection(
                        "jdbc:mysql://localhost:3306/testDB?useUnicode=
                            true&characterEncoding=UTF-8","root", null);
            stmt = conn.createStatement();
            rs = stmt.executeQuery(sql);
        } catch (SQLException ex) {
```

```java
            System.err.println(ex.getMessage());
        }
        return rs;
    }
    /**
     * 执行更新操作：insert、update、delete
     * */
    public int executeUpdate(String sql) {
        int result = 0;
        try {
            conn = DriverManager
                    .getConnection(
                            "jdbc:mysql://localhost:3306/testDB?useUnicode=" +
                                    "true&characterEncoding=UTF-8","root", null);
            stmt = conn.createStatement(ResultSet.TYPE_SCROLL_INSENSITIVE,
                    ResultSet.CONCUR_READ_ONLY);
            result = stmt.executeUpdate(sql);
        } catch (SQLException ex) {
            result = 0;
        }
        return result;
    }

    /**
     * 关闭数据库连接
     * */
    public void close() {
        try {
            if(rs != null)
                rs.close();
        } catch (Exception e) {
            e.printStackTrace(System.err);
        }
        try {
            if(stmt != null)
                stmt.close();
        } catch (Exception e) {
            e.printStackTrace(System.err);
        }
        try {
            if(conn != null) {
                conn.close();
            }
        } catch (Exception e) {
            e.printStackTrace(System.err);
        }
    }
}
```

分析连接数据库代码：

```
jdbc:mysql://localhost:3306/testDB?useUnicode=true&characterEncoding=UTF-8","root", null
```

可知：端口号为 3306，数据库名称为 testDB，数据库登录用户名为 root，密码为空。

第五步：新建待连接测试的数据库及数据表，并在数据表中添加若干初始数据，如图 6-7~图 6-10 所示。

图 6-7 新建待连接的数据库 testDB

图 6-8 数据库 testDB 创建成功

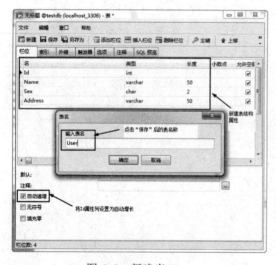

图 6-9 新建表 User

图 6-10 在表 User 中添加若干初始信息

第六步：测试是否连接成功。新建一个 testConnMysql.jsp 页面。

testConnMysql.jsp 页面代码如下：

```
<!--凡是涉及数据的操作请一定要导入java.sql.*-->
<%@ page language="java" import="java.util.*,java.sql.*" pageEncoding="utf-8"%>
<!--导入连接Mysql数据库的javabean-->
<jsp:useBean id="c" scope="page" class="com.ch6.Conn"/>
```

第 6 章 JDBC 技术详解

```
<!--实现数据查询,如果有数据显示在此网页中,证明连接成功,否则连接不成功。前提条件:
数据库testDB中的user表有记录-->

<%
String SQLSTR ="SELECT * FROM USER ORDER BY id DESC";
ResultSet RS = c.executeQuery(SQLSTR);  //执行查询
while(RS.next())
{
  //输出USER表中的各属性列中的内容
  out.print(RS.getString("name")+"\t");
  out.print(RS.getString("sex")+"\t");
  out.print(RS.getString("address")+"\t");
  out.print("<br>");
}
c.close();  //释放连接
%>
```

在浏览器中输入连接测试地址 http://localhost:8080/connMysql/testConnMysql.jsp,如果能将数据库 testDB 中的 User 表中的数据输出在 testConnMysql.jsp 文件中,则说明使用 JDBC 连接 MySQL 数据库成功。显示结果如图 6-11 所示。

图 6-11 测试是否连接成功

6.2.2 连接 SQL Server 2000 数据库

第一步:安装 SQL Server 2000 的 JDBC 驱动,驱动压缩包名称如图 6-12 所示。假设安装路径为 C:\Program Files\Microsoft SQL Server 2000 Driver for JDBC。

第二步:修改环境变量 Classpath,在变量值添加以下代码:

```
; C:\Program Files\Microsoft SQL Server 2000 Driver for
JDBC\lib\msbase.jar; C:\Program Files\Microsoft SQL Server 2000 Driver for
JDBC\lib\mssqlserver.jar; C:\Program Files\Microsoft SQL Server 2000 Driver
for JDBC\lib\msutil.jar
```

> **注意**
> 在最前面有一个分号。

第三步:复制 msbase.jar、mssqlserver.jar、msutil.jar 包到项目的 WEB-INF\lib 目录下,这个驱动的 JAR 包如图 6-13 所示。

图 6-12　SQL Serveer 2000 驱动压缩包

图 6-13　SQL Server 2000 驱动程序

以上驱动位于安装目录 C:\Program Files\Microsoft SQL Server 2000 Driver for JDBC\lib 下。

第四步：新建 SQL Server 2000 数据库的 JavaBean 代码如下：

```java
package com.ch6;
import java.sql.*;
public class connSqlserver2000 {
    Connection conn=null;
    Statement stmt=null;
    ResultSet rs=null;
    /**
     * 装载连接SQL Server 2000的连接驱动
     */
    public connSqlserver2000() {
        try {
            Class.forName("com.microsoft.jdbc.sqlserver.SQLServerDriver");
        } catch (java.lang.ClassNotFoundException e) {
            System.err.println(e.getMessage());
        }
    }
    /**
     * 执行查询
     * @param sql
     * @return
     */
    public ResultSet executeQuery(String sql) {
        try {
            conn=DriverManager
                    .getConnection("jdbc:microsoft:sqlserver://localhost:1433;DatabaseName=testDB;user=sa;password=1234");
            stmt=conn.createStatement(ResultSet.TYPE_SCROLL_INSENSITIVE,
                    ResultSet.CONCUR_READ_ONLY);
            rs=stmt.executeQuery(sql);
        } catch (SQLException ex) {
            System.err.println(ex.getMessage());
        }
        return rs;
    }
    /**
     * 执行更新
     * @param sql
     * @return
```

```java
    */
    public int executeUpdate(String sql) {
        int result=0;
        try {
            conn=DriverManager
                    .getConnection("jdbc:microsoft:sqlserver://localhost:1433;DatabaseName=testDB;user=sa;password=1234");
            stmt=conn.createStatement(ResultSet.TYPE_SCROLL_INSENSITIVE,
                    ResultSet.CONCUR_READ_ONLY);
            result=stmt.executeUpdate(sql);
        } catch (SQLException ex) {
            result=0;
        }
        return result;
    }
    /**
     * 关闭数据库连接
     */
    public void close() {
        try {
            if(rs != null)
                rs.close();
        } catch (Exception e) {
            e.printStackTrace(System.err);
        }
        try {
            if(stmt != null)
                stmt.close();
        } catch (Exception e) {
            e.printStackTrace(System.err);
        }
        try {
            if(conn != null) {
                conn.close();
            }
        } catch (Exception e) {
            e.printStackTrace(System.err);
        }
    }
}
```

测试连接 SQL Server 2000 数据库的方式与 6.2.1 中连接测试 MySQL 一致，此处不再详细解释。

6.2.3 连接 SQL Server 2005 数据库

第一步：解压 SQL Server 2005 的 JDBC 驱动，安装包名称如图 6-14 所示。

第二步：复制 sqljdbc.jar 到项目的 WEB-INF\lib 目录下，这个驱动的 JAR 包如图 6-15 所示。

图 6-14 SQL Server 2005 驱动安装包

图 6-15 SQL Server 2005 jar 文件

第三步：连接到 SQL Server 2005 数据库的 JavaBean。代码如下：

```java
package com.ch6;
import java.sql.*;
public class connSqlserver2005 {
    Connection conn = null;
    Statement stmt = null;
    ResultSet rs = null;
    /**
     * 装载连接SQL Server 2005驱动
     */
    public connSqlserver2005() {
        try {
            Class.forName("com.microsoft.sqlserver.jdbc.SQLServerDriver");
        } catch (java.lang.ClassNotFoundException e) {
            System.err.println(e.getMessage());
        }
    }
    /**
     * 执行查询
     *
     * @param sql
     * @return
     */
    public ResultSet executeQuery(String sql) {
        try {
            conn = DriverManager.getConnection(
                    "jdbc:sqlserver://localhost:1951;DatabaseName=testDB",
                    "sa", "1234");
            stmt = conn.createStatement(ResultSet.TYPE_SCROLL_INSENSITIVE,
                    ResultSet.CONCUR_READ_ONLY);
            rs = stmt.executeQuery(sql);
        } catch (SQLException ex) {
            System.err.println(ex.getMessage());
        }
        return rs;
    }
    /**
     * 执行更新
     *
     * @param sql
```

```java
 * @return
 */
public int executeUpdate(String sql) {
    int result = 0;
    try {
        conn = DriverManager.getConnection(
                "jdbc:sqlserver://localhost:1951;DatabaseName=testDB",
                "sa", "1234");
        stmt = conn.createStatement(ResultSet.TYPE_SCROLL_INSENSITIVE,
                ResultSet.CONCUR_READ_ONLY);
        result = stmt.executeUpdate(sql);
    } catch (SQLException ex) {
        result = 0;
    }
    return result;
}
/**
 * 关闭数据库连接
 */
public void close() {
    try {
        if(rs != null)
            rs.close();
    } catch (Exception e) {
        e.printStackTrace(System.err);
    }
    try {
        if(stmt != null)
            stmt.close();
    } catch (Exception e) {
        e.printStackTrace(System.err);
    }
    try {
        if(conn != null) {
            conn.close();
        }
    } catch (Exception e) {
        e.printStackTrace(System.err);
    }
}
}
```

连接 SQL Server 2005 时，注意要将 SQL Server 2005 的端口开启。依次选择"开始"→ "Microsoft SQL Server 2005" → "配置工具" → "SQL Server Configuration Manager" 命令，如图 6-16 所示。单击进入图 6-17 所示界面。在图 6-17 中的 TCP/IP 上右击，选择"属性"命令，出现如图 6-18 所示界面。输入 SQL Server 2005 的 TCP/IP 端口。

图 6-16 选择 SQL Server Configuration Manager 命令

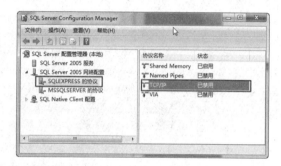

图 6-17 SQL Server Configuration Manager 窗口

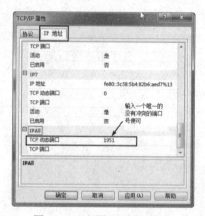

图 6-18 配置 TCP/IP 端口

6.3 JSP 操作 MySQL 数据库

本小节主要讲解 JSP 实现对 MySQL 数据库的四个基本操作,为保证实例的整体性,我们将以一个简单的学生信息管理为实例,实现数据库的查询、修改、添加、删除功能,如图 6-19 所示。

数据库设计是项目开发中非常重要的一个环节,它就像高楼大厦的根基一样,如果建设不好,在后来的系统维护、变更和功能扩充时,甚至在系统开发过程中,将会引起比较大的问题,会遇到非常大的困难。在此小节中,为让初学者更容

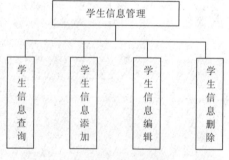

图 6-19 功能模块

易理解 JSP 操作数据库的过程,所以只选择一个数据关系表进行操作,学生信息表的描述如表 6-1 所示。在 Navicat 中创建 student 表,如图 6-20 所示。

第 6 章　JDBC 技术详解

表 6-1　学生信息描述表 student

字 段 名	类　　型	长　　度	是否允许为空	是否主键	描　　述
Id	int	4	否	是	自动增长编号
Number	varchar	16	否	否	学号
Name	varchar	25	是	否	姓名
Sex	char	2	是	否	性别
Address	varchar	50	是	否	住址
Phone	varchar	25	是	否	联系电话
Email	varchar	25	是	否	邮箱地址

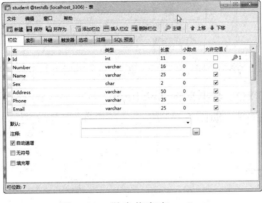

图 6-20　学生信息表 student

6.3.1　数据查询

第一步：新建 Web Project，如图 6-21 所示。

图 6-21　新建工程 DBControl

第二步：把 MySQL 的 JDBC 驱动程序复制到项目的 WEB-INF\lib 目录下，这个驱动的 JAR 包为 mysql-connector-java-5.1.6-bin.jar。驱动复制成功后，工程目录结构如图 6-22 所示。

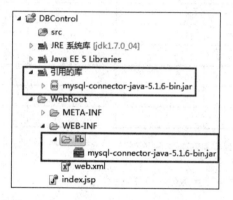

图 6-22 粘贴 mysql-connector-java-5.1.6-bin.jar 后工程目录结构

第三步：创建连接到 MySQL 数据库的 JavaBean。在工程中的 src 目录上右击，选择"新建"→"类"命令，如图 6-23 所示，在"新建 Java 类"对话框中进行设置，如图 6-24 所示。

图 6-23 选择"新建"→"类"命令

图 6-24 "新建 Java"对话框

第四步：在工程 DBControl 的 com.ch6.Conn.java 中输入代码，实现连接 MySQL 数据库。代码如下：

```
package com.ch6;
import java.sql.*;
public class Conn {
    Connection conn = null;
```

第6章 JDBC 技术详解

```java
        Statement stmt = null;
        ResultSet rs = null;
    /**
     * 加载驱动程序
     * */
    public Conn() {
        try {
            Class.forName("com.mysql.jdbc.Driver");
        } catch (java.lang.ClassNotFoundException e) {
            System.err.println(e.getMessage());
        }
    }
    /**
     * 执行查询操作: select
     * */
    public ResultSet executeQuery(String sql) {
        try {
            conn = DriverManager
                    .getConnection(
                        "jdbc:mysql://localhost:3306/testDB?useUnicode=true&characterEncoding=UTF-8","root", null);
            stmt = conn.createStatement();
            rs = stmt.executeQuery(sql);
        } catch (SQLException ex) {
            System.err.println(ex.getMessage());
        }
        return rs;
    }
    /**
     * 执行更新操作: insert、update、delete
     * */
    public int executeUpdate(String sql) {
        int result = 0;
        try {
            conn = DriverManager
                    .getConnection(
                        "jdbc:mysql://localhost:3306/testDB?useUnicode=true&characterEncoding=UTF-8","root", null);
            stmt = conn.createStatement(ResultSet.TYPE_SCROLL_INSENSITIVE,
                    ResultSet.CONCUR_READ_ONLY);
            result = stmt.executeUpdate(sql);
        } catch (SQLException ex) {
            result = 0;
        }
        return result;
    }
    /**
     * 关闭数据库连接
     * */
    public void close() {
```

```java
        try {
            if(rs != null)
                rs.close();
        } catch (Exception e) {
            e.printStackTrace(System.err);
        }
        try {
            if(stmt != null)
                stmt.close();
        } catch (Exception e) {
            e.printStackTrace(System.err);
        }
        try {
            if(conn != null) {
                conn.close();
            }
        } catch (Exception e) {
            e.printStackTrace(System.err);
        }
    }
}
```

分析 com.ch6.Conn.java 代码可知：连接的数据库为 MySQL，数据库名称为 testDB，数据库登录用户名为 root，密码为空。

第五步：创建一个模型 com.ch6.model.StudentInfo.java，该模型只有一些属性及其 getter 与 setter 方法的类，此类没有业务逻辑。代码如下：

```java
package com.ch6.model;

public class StudentInfo {
    private int id;
    private String number;
    private String name;
    private String sex;
    private String address;
    private String phone;
    private String email;

    public int getId() {
        return id;
    }
    public void setId(int id) {
        this.id=id;
    }
    public String getNumber() {
        return number;
    }
    public void setNumber(String number) {
        this.number=number;
```

```java
    }
    public String getName() {
        return name;
    }
    public void setName(String name) {
        this.name=name;
    }
    public String getSex() {
        return sex;
    }
    public void setSex(String sex) {
        this.sex=sex;
    }
    public String getAddress() {
        return address;
    }
    public void setAddress(String address) {
        this.address=address;
    }
    public String getPhone() {
        return phone;
    }
    public void setPhone(String phone) {
        this.phone=phone;
    }
    public String getEmail() {
        return email;
    }
    public void setEmail(String email) {
        this.email=email;
    }
}
```

第六步：新建一个进行数据库操作的类 com.ch6.dal.Student.java。代码如下：

```java
package com.ch6.dal;
import java.sql.ResultSet;
import java.sql.SQLException;
import java.util.ArrayList;
import java.util.List;
import com.ch6.Conn;
import com.ch6.model.StudentInfo;

public class Student {
    Conn conn=new Conn();
    /**
     * 获取学生列表
     * @return
     * @throws SQLException
```

```java
    */
    public List<StudentInfo> getList() throws SQLException {
        List<StudentInfo> list=new ArrayList<StudentInfo>();
        String sql= "select* * from student order by number asc";
        ResultSet rs=conn.executeQuery(sql);
        while (rs.next()) {
            StudentInfo info=new StudentInfo();
            info.setId(rs.getInt("Id"));
            info.setNumber(rs.getString("Number"));
            info.setName(rs.getString("Name"));
            info.setAddress(rs.getString("Address"));
            info.setPhone(rs.getString("Phone"));
            info.setSex(rs.getString("Sex"));
            info.setEmail(rs.getString("Email"));
            list.add(info);
        }
        conn.close();
        return list;
    }
}
```

第七步：在 WebRoot 目录下新建一个 select.jsp 页面，如图 6-25 所示。实现数据查询（列表显示）。

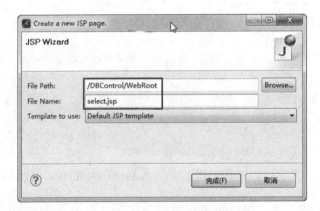

图 6-25 新建 select.jsp 页面

select.jsp 页面代码如下：

```jsp
<%@ page language="java" import="java.util.*" pageEncoding="utf-8"%>
<%@page import="com.ch6.model.StudentInfo,com.ch6.dal.Student"%>



<!DOCTYPE HTML PUBLIC "-//W3C//DTD HTML 4.01 Transitional//EN">
<html>
  <head>
    <title>数据查询</title>
  </head>
  <body>
```

```
    <table width="539" border="1">
    <tr>
     <td>学号</td>
     <td>姓名</td>
     <td>性别</td>
     <td>住址</td>
     <td>电话</td>
     <td>邮箱</td>
     <td>操作</td>
    </tr>
    <%
      Student student = new Student(); //创建 com.ch6.dal.Student 的对象,命
名为 student
      List<StudentInfo> list = student.getList();//通过 student 对象调用方法
getList()获取学生列表信息,该方法返回一个 List 集合
      for(StudentInfo info:list) {   //遍历输出 list 集合中的数据
    %>
    <tr>
     <td><%out.print(info.getNumber());//调用对象的属性值 %></td>
     <td><%out.print(info.getName()); %></td>
     <td><%out.print(info.getSex()); %></td>
     <td><%out.print(info.getAddress()); %></td>
     <td><%out.print(info.getPhone()); %></td>
     <td><%out.print(info.getEmail()); %></td>
     <td><a href=" ">编辑</a> | <a href=" ">删除</a></td>
    </tr>
    <%} %>
    </table>
    </body>
    </html>
```

第八步：在 MyEclipse 中部署该工程，启动 Tomcat 8，在浏览器中输入文件路径，预览效果如图 6-26 所示。测试 URL：http://localhost:8080/DBControl/select.jsp。

图 6-26　浏览查询结果

第九步：设计一个导航页 menu.jsp，页面布局如图 6-27 所示。

图 6-27 menu.jsp 页面布局

```
<%@ page language="java" import="java.util.*" pageEncoding="utf-8"%>
<table width="386" border="1">
  <tr>
    <td colspan="4">学生信息管理</td>
  </tr>
  <tr>
    <td><a href="select.jsp">查询学生</a></td>
    <td><a href="insert.jsp">添加学生</a></td>
  </tr>
</table>
```

为提高该实例的界面友好性，设计一个框架，将添加、编辑、删除、查询连接起来。

第十步：使用 Dreamweaver 创建一个上下结构的框架，如图 6-28～图 6-32 所示。

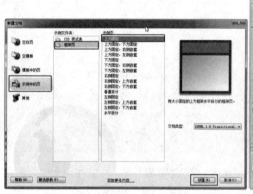

图 6-28 创建上下结构的框架

图 6-29 指定框架标题

图 6-30 设置框架页高度

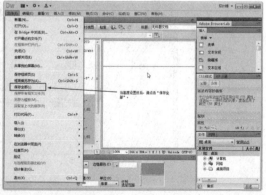

图 6-31 保存框架集

图 6-32 保存框架集到项目 WebRoot 目录

该框架集页面 index.htm 代码如下：

```
<!DOCTYPE html PUBLIC "-//W3C//DTD XHTML 1.0 Frameset//EN"
"http://www.w3.org/TR/xhtml1/DTD/xhtml1-frameset.dtd">
<html xmlns="http://www.w3.org/1999/xhtml">
<head>
<meta http-equiv="Content-Type" content="text/html; charset=utf-8" />
<title>学生管理系统</title>
</head>
<frameset rows="97,*" cols="*" framespacing="0" frameborder="no" border="0">
  <!--上半部分默认命名为topFrame-->
  <frame src="menu.jsp" name="topFrame" scrolling="no" noresize="noresize" id="topFrame" title="topFrame" />
  <!--下半部分默认命名为mainFrame-->
  <frame src="select.jsp" name="mainFrame" id="mainFrame" title="mainFrame" />
</frameset>
<noframes><body>
</body></noframes>
</html>
```

第十一步：刷新工程。因为该 index.htm 不是在 MyEclipse 软件中创建保存的，所以要刷新该工程文件，让部署目录与工程文件目录同步，如图 6-33 所示。在浏览器中输入文件路径 http://localhost:8080/DBControl/index.htm，预览结果如图 6-34 所示。

第十二步：在 index.htm 框架集页面，默认执行 menu.jsp 与 select.jsp 两个文件。当单击

menu.jsp 文件中的"查询学生"链接，应当在框架集合的下半部分显示（在创建 index.htm 时，下半部分框架页命名为 mainFrame），为此要修改 menu.jsp 文件中的两行代码，增加一个跳转目录属性，即 target="mainFrame"，修改后的 menu.jsp 代码如下所示。

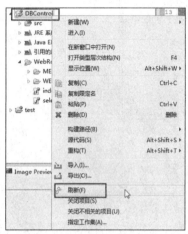

图 6-33　刷新工程文件

图 6-34　浏览整个框架页面

```
<%@ page language="java" import="java.util.*" pageEncoding="utf-8"%>
<table width="386" border="1">
  <tr>
    <td colspan="4">学生信息管理</td>
  </tr>
  <tr>
    <td><a href="select.jsp" target="mainframe">查询学生</a></td>
    <td><a href="insert.jsp" target="mainFrame">添加学生</a></td>
  </tr>
</table>
```

6.3.2　数据添加

第一步：新建 insert.jsp 保存为工程目录 WebRoot 下。页面布局与代码如图 6-35 所示。

图 6-35　insert.jsp 页面布局

```
<%@ page language="java" import="java.util.*" pageEncoding="utf-8"%>
<!DOCTYPE HTML PUBLIC "-//W3C//DTD HTML 4.01 Transitional//EN"
```

```html
<html>
  <head>
    <title>数据添加</title>
  </head>
  <body>
    <form id="form1" name="form1" method="post" action="">
    <table width="384" height="289" border="1">
    <tr>
      <td>学号</td>
      <td><input type="text" name="number" id="number" /></td>
    </tr>
    <tr>
      <td>姓名</td>
      <td><input type="text" name="name" id="name" /></td>
    </tr>
    <tr>
      <td>性别</td>
      <td>
        <input type="radio" name="sex" id="sex" value="男" checked/> 男
        <input type="radio" name="sex" id="sex" value="女" /> 女
      </td>
    </tr>
    <tr>
      <td>住址</td>
      <td><input type="text" name="address" id="address"/></td>
    </tr>
    <tr>
      <td>电话</td>
      <td><input type="text" name="phone" id="phone" /></td>
    </tr>
    <tr>
      <td>邮箱</td>
      <td><input type="text" name="email" id="email" /></td>
    </tr>
    <tr>
      <td colspan="2"><input type="submit" name="button" id="button" value="提交" />
        <input type="reset" name="button2" id="button2" value="重置" />
      </td>
    </tr>
    </table>
    </form>
  </body>
</html>
```

第二步：当单击 insert.jsp 中的"提交"按钮时，要实现数据的保存。下面实现在 insert.jsp 文件中实现表间数据处理，将数据保存在数据库中。修改 insert.jsp 文件，修改后代码如下所示。

```jsp
<%@ page language="java" import="java.util.*" pageEncoding="utf-8"%>
```

```jsp
<%@page import="com.ch6.model.StudentInfo,com.ch6.dal.Student"%>
<%
request.setCharacterEncoding("utf-8"); //设置编辑，中文数据保存不会出现乱码
StudentInfo info=new StudentInfo(); //创建 com.ch6.model.StudentInfo 的对象 info
Student student = new Student(); //创建 com.ch6.dal.Student 的对象 student
//如果获得参数 action 的值为 add ，则表示，操作者已经单击了"提交"按钮，那么将执行下面代码实现数据保存
    if("add".equals(request.getParameter("action"))  )
    {
        info.setNumber(request.getParameter("number"));
        info.setName(request.getParameter("name"));
        info.setSex(request.getParameter("sex"));
        info.setAddress(request.getParameter("address"));
        info.setPhone(request.getParameter("phone"));
        info.setEmail(request.getParameter("email"));
        student.insert(info); //通过 student 对象调用 insert()方法，实现数据保存
    }
%>
<!DOCTYPE HTML PUBLIC "-//W3C//DTD HTML 4.01 Transitional//EN">
<html>
  <head>
    <title>数据添加</title>
  </head>
  <body>
    <!-- 注意 form 标签的 action 属性值   -->
    <form id="form1" name="form1" method="post" action="insert.jsp?action=add">
    <table width="384" height="289" border="1">
      <tr>
        <td>学号</td>
        <td><input type="text" name="number" id="number" /></td>
      </tr>
      <tr>
        <td>姓名</td>
        <td><input type="text" name="name" id="name" /></td>
      </tr>
      <tr>
        <td>性别</td>
        <td>
          <input type="radio" name="sex" id="sex" value="男" checked/> 男
          <input type="radio" name="sex" id="sex" value="女" /> 女
        </td>
      </tr>
      <tr>
        <td>住址</td>
        <td><input type="text" name="address" id="address"/></td>
      </tr>
      <tr>
        <td>电话</td>
        <td><input type="text" name="phone" id="phone" /></td>
      </tr>
      <tr>
        <td>邮箱</td>
        <td><input type="text" name="email" id="email" /></td>
```

```
        </tr>
        <tr>
          <td colspan="2"><input type="submit" name="button" id="button" value="提交" />
          <input type="reset" name="button2" id="button2" value="重置" />
          </td>
        </tr>
      </table>
    </form>
  </body>
</html>
```

第三步：在 com.ch6.dal.Student.java 中新增一个方法，实现数据保存。代码如下所示。

```
    …
    /**
     * 添加
     * @param info
     * @return
     */
    public int insert(StudentInfo info)
    {
        String sql = "insert into student(number,name,address,phone,sex,email) values ";
        sql = sql + " ('"+info.getNumber()+"','"+info.getName()+"','"+info.getAddress()+"','"+info.getPhone()+"','"+info.getSex()+"','"+info.getEmail()+"')";
        int result = 0;
        System.out.println(sql);
        result = conn.executeUpdate(sql);
        conn.close();
        return result;
    }
    …
```

第四步：刷新工程或者重新启动 Tomcat。在浏览器地址栏中输入 http://localhost:8080/DBControl/index.htm，如图 6-36 和图 6-37 所示。

图 6-36　测试添加学生

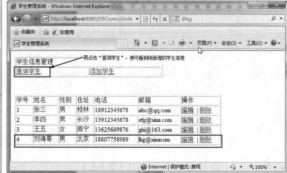
图 6-37　添加学生成功

打开 Navicat for MySQL，刷新数据表 student，检查数据是否保存的关系表中，如图 6-38 所示。

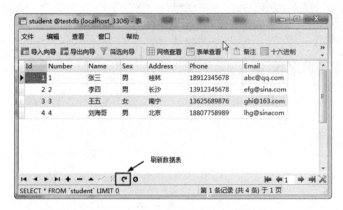

图 6-38　检查关系表中数据是否保存

6.3.3　数据编辑

数据的编辑思路：先找到数据，将数据显示在可编辑的文本框中，当数据修改好后，单击按钮，实现数据保存。

第一步：在 com.ch6.dal.Student.java 中新增两个方法，一个用于实现数据编辑，另一个用于查找当前编辑的学生信息。代码如下：

```java
/**
 * 编辑
 * @param info
 * @return
 */
public int update(StudentInfo info) {
    String sql = "update student set"
        + " number='"+info.getNumber()+"',name='"+info.getName()+"',address='"+info.getAddress()+"',phone='"+info.getPhone()+"',sex='"+info.getSex()+"',email='"+info.getEmail()+"' where id='"+info.getId()+"'";
    int result = 0;
    System.out.println(sql);
    result = conn.executeUpdate(sql);
    conn.close();
    return result;
}
/**
 * 获取单个学生信息
 * @param id
 * @return
 * @throws SQLException
 */
public StudentInfo getStudent(String id) throws SQLException{
    StudentInfo info= new StudentInfo();
    String sql ="select * from student s where id='"+id+"'";
    ResultSet rs = conn.executeQuery(sql);
    if(rs.next())
```

```
        {
            info.setId(rs.getInt("Id"));
            info.setNumber(rs.getString("Number"));
            info.setName(rs.getString("Name"));
            info.setAddress(rs.getString("Address"));
            info.setPhone(rs.getString("Phone"));
            info.setSex(rs.getString("Sex"));
            info.setEmail(rs.getString("Email"));
        }
        conn.close();
        return info;
    }
```

第二步：修改 select.jsp 页面的编辑按钮链接，传递一个唯一编号作为修改的条件（在设计 student 表时，将 id 字段设置成了自动增长编号，此属性列是主键）。修改后代码如下：

```
<a href="update.jsp?id=<%=info.getId()%>">编辑</a>
```

第三步：新建 update.jsp 页面（温馨提示：可将 insert.jsp 文件中的代码复制到 update.jsp 文件中，只需做简单的修改），代码如下所示。

```
<%@ page language="java" import="java.util.*" pageEncoding="utf-8"%>
<%@page import="com.ch6.model.StudentInfo,com.ch6.dal.Student"%>
<%
request.setCharacterEncoding("utf-8"); //设置编辑,中文数据保存不会出现乱码
StudentInfo info=new StudentInfo(); //创建 com.ch6.model.StudentInfo 的对
                                    //象 info
Student student = new Student(); //创建 com.ch6.dal.Student 的对象 student
//如果获得参数 action 的值为 add，则表示,操作者已经单击了"提交"按钮,那么将执行下
//面代码实现数据保存
    if("edit".equals(request.getParameter("action"))  )
    {
        info= student.getStudent(request.getParameter("id"));
        if(info==null)
        {
            out.print("找不到该学生信息");
        }
        info.setId(Integer.parseInt(request.getParameter("id")));
        info.setNumber(request.getParameter("number"));
        info.setName(request.getParameter("name"));
        info.setSex(request.getParameter("sex"));
        info.setAddress(request.getParameter("address"));
        info.setPhone(request.getParameter("phone"));
        info.setEmail(request.getParameter("email"));
        student.update(info);
    }
%>
<!DOCTYPE HTML PUBLIC "-//W3C//DTD HTML 4.01 Transitional//EN">
<html>
```

```jsp
<head>
    <title>数据编辑</title>
</head>
<body>
<%
StudentInfo sinfo = student.getStudent(request.getParameter("id"));
                                    //先找到该编号的学生信息
if(sinfo==null){
    out.print("找不到该学生信息");
    return;
}
%>
    <form id="form1" name="form1" method="post" action="update.jsp?action=edit">
    <table width="384" height="289" border="1">
    <tr>
        <td>学号</td>
        <td><input type="text" name="number" id="number" value="<%=sinfo.getNumber() %>"/></td>
    </tr>
    <tr>
        <td>姓名</td>
        <td><input type="text" name="name" id="name" value="<%=sinfo.getName() %>"/></td>
    </tr>
    <tr>
        <td>性别</td>
        <td><input type="radio" name="sex" id="sex" value="男" <%if("男".equals(sinfo.getSex())) out.print("checked"); %>/>
        男
        <input type="radio" name="sex" id="sex" value="女" <%if("女".equals(sinfo.getSex())) out.print("checked"); %>/>
        女</td>
    </tr>
    <tr>
        <td>住址</td>
        <td><input type="text" name="address" id="address" value="<%=sinfo.getAddress() %>"/></td>
    </tr>
    <tr>
        <td>电话</td>
        <td><input type="text" name="phone" id="phone" value="<%=sinfo.getPhone() %>"/></td>
    </tr>
    <tr>
        <td>邮箱</td>
        <td><input type="text" name="email" id="email" value="<%=sinfo.getEmail() %>"/></td>
    </tr>
    <tr>
```

第 6 章　JDBC 技术详解

```html
      <td colspan="2"><input type="submit" name="button" id="button" value="提交" />
      <input type="reset" name="button2" id="button2" value="重置" />
      <!--此处传递一当前要编辑的学生信息id值，此id值不允许修改，设置为隐藏域 -->
      <input type="hidden" name="id" value="<%= sinfo.getId() %>" />
      </td>
    </tr>
  </table>
</form>
  </body>
</html>
```

第四步：刷新工程或者重新启动 Tomcat，在浏览器地址栏中输入 http://localhost:8080/DBControl/index.htm，进行修改操作，如图 6-39～图 6-41 所示。

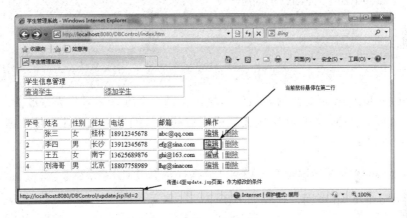

图 6-39　测试数据编辑

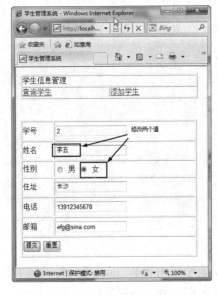

图 6-40　修改若干项

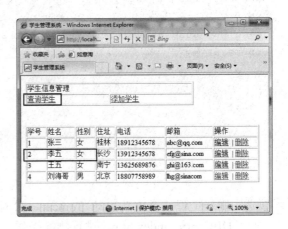

图 6-41　修改成功

-107-

6.3.4 数据删除

第一步：在 com.ch6.dal.Student.java 中新增一个方法，实现数据编辑。代码如下：

```java
/**
 * 删除
 * @param id
 * @return
 */
public int delete(String id)
{
    String sql = "delete from student where id ='"+id+"'";
    int result = 0;
    System.out.println(sql);
    result = conn.executeUpdate(sql);
    conn.close();
    return result;
}
```

第二步：修改 select.jsp 页面的编辑按钮链接，传递一个唯一编号作为修改的条件（在设计 student 表时，将 id 字段设置成了自动增长编号，此属性列是主键）。修改后代码如下：

```html
<a href="delete.jsp?id=<%=info.getId()%>">删除</a>
```

第三步：新建 delete.jsp 页面，代码如下所示。

```jsp
<%@ page language="java" import="java.util.*" pageEncoding="utf-8"%>
<%@page import="com.ch6.dal.Student"%>

<%
Student student= new Student(); //创建对象
int result = 0;
result = student.delete(request.getParameter("id"));
if(result==1)
{
  out.print("删除成功");
}else
{
  out.print("删除失败");
}
%>
```

第四步：刷新工程或者重新启动 Tomcat。在浏览器地址栏中输入 http://localhost:8080/DBControl/index.htm，进行删除操作，如图 6-42～图 6-44 所示。

图 6-42　测试删除数据

图 6-43　删除数据成功

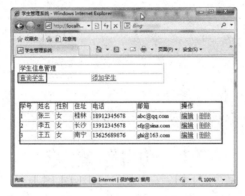

图 6-44　查询学生找不到刚删除的学生信息

小　　结

数据库编程是 JSP 学习的难点，也是重点。只有掌握了数据库编程知识，JSP 知识才有用武之地。几乎所有的 JSP 项目均使用到数据库，可见数据库编程的重要性。

在 Java 中，数据库编程通过 JDBC 实现。通过 JDBC 操作数据库的大致过程如下：第一步：注册并加载驱动程序。第二步：创建连接。第三步：创建 SQL 语句对象。第四步：提交 SQL 语句。第五步：显示结果。第六步：关闭连接。

习　　题

新建一个工程 Ex6，在本工程中实现如图 6-45 所示的功能。

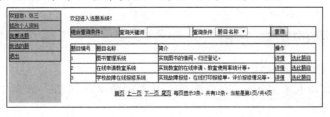

图 6-45　Ex6 功能界面

第 7 章 综合实例——BLOG 系统开发

学习目标
- JSP 开发。
- JavaBean 开发。

本章我们将采用 JSP+JavaBean 来开发一个博客系统。

7.1 功能要求

根据博客系统的基本要求，本系统需要完成如下的主要任务。

（1）系统分前台和后台，后台管理员登录成功后，可以发布博文，可以对博文、博文分类、博文评论、注册用户等进行管理，管理员可以修改自己的登录密码。

（2）浏览者进入博客前台后，能浏览里面的文章。但是必须是注册用户才可以回复。

（3）能按分类进行博文搜索。

（4）前台会员注册信息包括：用户名、密码、E-mail。

（5）密码要进行加密处理。

（6）注册用户登录成功后，能进行密码修改，能查看自己评论过的文章。

7.2 数据库设计

应用系统的开发几乎离不开数据库，很多应用程序都包括了对数据库的数据检索、更新、插入、删除等操作。因此，一个简单易用并且功能强大的数据库系统就成为应用软件开发不可或缺的一部分。

7.2.1 数据库的需求分析

博客系统的数据库表设计如下：

（1）博文发布信息表（blog）：存放博主发布的博文，比如：博文标题、内容、发布时间、所属分类等。

（2）博文分类信息表（class）：存放博文分类信息，比如：博文分类编号、博文分类名称、排序等。

（3）博文评论信息表（comment）：存放博文评论信息，比如：博文评论内容、评论时间、评论人、评论的博文等。

（4）用户信息表（users）：存放用户信息，比如：用户账号、密码、E-mail、身份等。

7.2.2 数据库的逻辑设计

本系统采用 MySQL 作为后台数据库进行开发，数据库名称为 blogdb。

各关系表的字段描述信息如表 7-1～表 7-4 所示。

表 7-1 博文发布信息表 blog

字 段 名	类 型	长 度	是否允许为空	是否主键	描 述
Id	int	4	否	是	自动增长编号
Title	varchar	255	否	否	博文标题
Context	text		是	否	博文内容
CreatedTime	datetime	8	是	否	发布时间
ClassId	int	4	否	否	所属分类编号

表 7-2 博文分类信息表 class

字 段 名	类 型	长 度	是否允许为空	是否主键	描 述
Id	int	4	否	是	自动增长编号（博文分类编号）
Name	varchar	50	否	否	分类名称
Sort	int	4	否	否	排序

表 7-3 博文评论信息表 comment

字 段 名	类 型	长 度	是否允许为空	是否主键	描 述
Id	int	4	否	是	自动增长编号
Context	text		是	否	评论内容
CreatedTime	datetime	4	否	否	评论时间
UserName	varchar	50	否	否	评论人
BlogId	int	4	否	否	评论的博文编号

表 7-4 用户信息表 users

字 段 名	类 型	长 度	是否允许为空	是否主键	描 述
UserName	varchar	16	否	是	用户名
Password	varchar	16	否	否	密码
Email	varchar	50	是	否	邮箱
Power	varchar	50	否	否	身份：admin－表示管理员；user－表示一般注册用户

数据库和各关系表的具体实现如图 7-1～图 7-5 所示。

图 7-1 创建数据库 blogdb

图 7-2 创建 blog 信息表

图 7-3 创建 class 信息表　　　　　　图 7-4 创建 comment 信息表

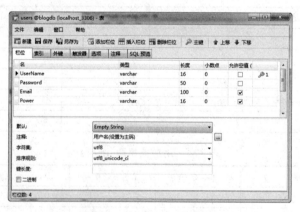

图 7-5 创建 users 信息表

7.3 框架搭建

第一步：启动 MyEclipse，新建一个名称为 blog 的 Web 工程文件，如图 7-6 所示。

第 7 章　综合实例——BLOG 系统开发

图 7-6　创建 Web 工程

第二步：在工程目录中的 src 目录下新建三个包，将 MySQL 数据库连接驱动复制至本工程目录的 lib 文件夹下，目录结构如图 7-7 和图 7-8 所示。

图 7-7　新建包

图 7-8　工程目录结构

第三步：在包 com.ch7.model 下分别新建 BlogInfo.java、ClassInfo.java、CommentInfo.java、UsersInfo.java 四个只有 get() 与 set() 方法的 JavaBean，这四个 JavaBean 的主要是进行值对象在 Servlet 与 JSP 之间传递数据，如图 7-9 所示。

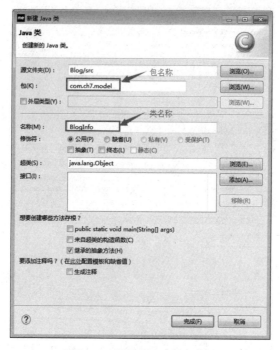

图 7-9 创建 BlogInfo.java

BlogInfo.java 的代码如下：

```java
package com.ch7.model;
import java.util.Date;
public class BlogInfo {
    private int id;
    private String title="";
    private String context="";
    private Date createdtime;
    private int classid;
    private String className="";

    public String getClassName() {
        return className;
    }
    public void setClassName(String className) {
        this.className = className;
    }
    public int getId() {
        return id;
    }
    public void setId(int id) {
        this.id = id;
    }
    public String getTitle() {
        return title;
    }
```

```java
    public void setTitle(String title) {
        this.title = title;
    }
    public String getContext() {
        return context;
    }
    public void setContext(String context) {
        this.context = context;
    }
    public Date getCreatedtime() {
        return createdtime;
    }
    public void setCreatedtime(Date createdtime) {
        this.createdtime = createdtime;
    }
    public int getClassid() {
        return classid;
    }
    public void setClassid(int classid) {
        this.classid = classid;
    }
}
```

ClassInfo.java 的代码如下:

```java
package com.ch7.model;

public class ClassInfo {
    private int id;
    private String name="";
    private int sort;
    public int getId() {
        return id;
    }
    public void setId(int id) {
        this.id = id;
    }
    public String getName() {
        return name;
    }
    public void setName(String name) {
        this.name = name;
    }
    public int getSort() {
        return sort;
    }
    public void setSort(int sort) {
        this.sort = sort;
    }
}
```

CommentInfo.java 的代码为：

```java
package com.ch7.model;

import java.util.Date;

public class CommentInfo {
    private int id;
    private String context="";
    private Date createdtime;
    private String username="";
    private int blogid;
    public int getId() {
        return id;
    }
    public void setId(int id) {
        this.id = id;
    }
    public String getContext() {
        return context;
    }
    public void setContext(String context) {
        this.context = context;
    }
    public Date getCreatedtime() {
        return createdtime;
    }
    public void setCreatedtime(Date createdtime) {
        this.createdtime = createdtime;
    }
    public String getUsername() {
        return username;
    }
    public void setUsername(String username) {
        this.username = username;
    }
    public int getBlogid() {
        return blogid;
    }
    public void setBlogid(int blogid) {
        this.blogid = blogid;
    }
}
```

UsersInfo.java 的代码如下：

```java
package com.ch7.model;

public class UsersInfo {
    private String username="";
    private String password="";
```

```java
    private String email="";
    private String power="";
    public String getUsername() {
        return username;
    }
    public void setUsername(String username) {
        this.username = username;
    }
    public String getPassword() {
        return password;
    }
    public void setPassword(String password) {
        this.password = password;
    }
    public String getEmail() {
        return email;
    }
    public void setEmail(String email) {
        this.email = email;
    }
    public String getPower() {
        return power;
    }
    public void setPower(String power) {
        this.power = power;
    }
}
```

第四步：在包 com.ch7.dal 目录下新建四个空的类 Blog.java（见图 7-10）、Comment.java、Class.java、Users.java 类。这四个类主要用于实现数据库的操作。

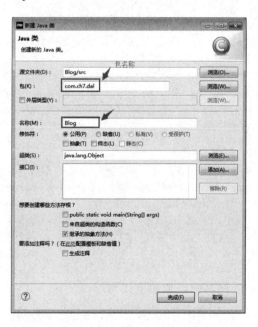

图 7-10　创建 Blog.java

第五步：在包 com.ch7.common 目录下新建几个通用的类：Utility.java（基类）、Data Converter.java（数据类型转换类）、DataValidator.java（数据验证类）、MD5.java（加密类）、Conn.java（数据库连接类）。这些类主要用于一些通用基本方法的实现。

第六步：本系统将使用到一个基于 Web 的文档在线编辑器——FCKEditor，下面将在本工程中配置 FCKEditor。

FCK 编辑器的使用与配置：

（1）在 WebRoot 目录下新建一个文件夹，命名为 fckeditor，将图 7-11 中的文件复制到此文件夹下。

（2）将 FCK 用到的相关 JAR 包复制到 WEB-INF/lib 目录下，如图 7-12 所示。

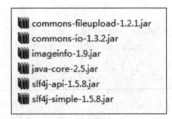

图 7-11　fckeditor 文件内容　　　　图 7-12　fckeditor 包

（3）在 src 根目录添加一个属性文件 fckeditor.properties 。该文件代码如下：

```
connector.userActionImpl=net.fckeditor.requestcycle.impl.EnabledUserAction
```

（4）配置 web.xml 文件。打开 WebRoot/WEB-INF/web.xml，在<web-app>、</web-app>标签中添加如下代码：

```xml
<!-- fckeditor -->
    <display-name>FCKeditor.Java Sample Web Application</display-name>
    <description>FCKeditor.Java Sample Web Application</description>
    <servlet>
        <servlet-name>ConnectorServlet</servlet-name>
        <servlet-class>
            net.fckeditor.connector.ConnectorServlet
        </servlet-class>
        <load-on-startup>1</load-on-startup>
    </servlet>
    <servlet-mapping>
        <servlet-name>ConnectorServlet</servlet-name>
        <!-- Do not wrap this line otherwise Glassfish will fail to load this file -->
        <url-pattern>/fckeditor/editor/filemanager/connectors/*</url-pattern>
    </servlet-mapping>
<!-- fckeditor -->
```

（5）在要使用 FCK 的网页中添加如下代码：

```
<%@page import="net.fckeditor.*"%>
…
```

```
<script type="text/javascript">
    function FCKeditor_OnComplete(editorInstance) {
        window.status = editorInstance.Description;
    }
</script>
…
  <%FCKeditor fckEditor = new FCKeditor(request, "content");
    fckEditor.setHeight("400"); %>
…
      <%= fckEditor %>
```

通过上面的几步操作，整个 blog 工程目录结构如图 7-13 所示。

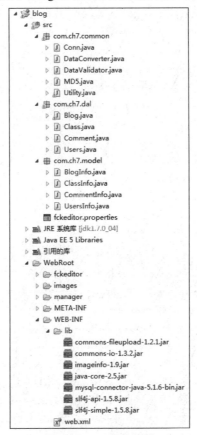

图 7-13　工程结构图

7.4　功　能　实　现

7.4.1　通用功能实现

1．Conn.java 实现

Conn.java 位于本工程的 com.ch7.common 目录，主要实现数据库连接。Conn.java 实现代码如下：

基础篇

```java
package com.ch7.common;
import java.sql.*;
/**
 * 数据库连接类
 */
public class Conn {
    Connection conn = null;
    Statement stmt = null;
    ResultSet rs = null;
    /**
     * 加载驱动程序
     * */
    public Conn() {
        try {
            Class.forName("com.mysql.jdbc.Driver");
        } catch (java.lang.ClassNotFoundException e) {
            System.err.println(e.getMessage());
        }
    }
    /**
     * 执行查询操作: select
     * */
    public ResultSet executeQuery(String sql) {
        try {
            conn = DriverManager
                    .getConnection(
                            "jdbc:mysql://localhost:3306/blogdb?useUnicode=true&characterEncoding=UTF-8","root", null);
            stmt = conn.createStatement();
            rs = stmt.executeQuery(sql);
        } catch (SQLException ex) {
            System.err.println(ex.getMessage());
        }
        return rs;
    }
    /**
     * 执行更新操作: insert、update、delete
     * */
    public int executeUpdate(String sql) {
        int result = 0;
        try {
            conn = DriverManager
                    .getConnection(
                            "jdbc:mysql://localhost:3306/blogdb?useUnicode=true&characterEncoding=UTF-8","root", null);
            stmt = conn.createStatement(ResultSet.TYPE_SCROLL_INSENSITIVE,
                    ResultSet.CONCUR_READ_ONLY);
            result = stmt.executeUpdate(sql);
        } catch (SQLException ex) {
            result = 0;
```

第 7 章 综合实例——BLOG 系统开发

```java
        }
        return result;
    }

    /**
     * 关闭数据库连接
     **/
    public void close() {
        try {
            if (rs != null)
                rs.close();
        } catch (Exception e) {
            e.printStackTrace(System.err);
        }
        try {
            if (stmt != null)
                stmt.close();
        } catch (Exception e) {
            e.printStackTrace(System.err);
        }
        try {
            if (conn != null) {
                conn.close();
            }
        } catch (Exception e) {
            e.printStackTrace(System.err);
        }
    }
}
```

2. DataConverter.java 实现

DataConverter.java 位于本工程的 com.ch7.common 目录，主要实现一些常用数据类型的转换。DataConverter.java 实现代码如下：

```java
package com.ch7.common;
import java.text.ParseException;
import java.text.SimpleDateFormat;
import java.util.Date;
/**
 * 数据转换类
 */
public class DataConverter {
    /**
     * 将日期格式化为字符串
     * @param date
     * @return
     */
    public static String dataToString(Date date){
        return dataToString(date, "yyyy-MM-dd HH:mm:ss");
    }
```

```java
/**
 * 将日期格式化为字符串
 * @param date - 日期
 * @param formatType - 格式化方式
 * @return - 字符串
 */
public static String dataToString(Date date, String formatType){
    if(date == null){
        date = Utility.getNowDateTime();
    }
    SimpleDateFormat formatter = new SimpleDateFormat(formatType);
    return formatter.format(date);
}
/**
 * 将字符串转成日期(yyyy-MM-dd HH:mm:ss)
 * @param input - 日期字符串
 * @return
 */
public static Date toDate(String input){
    return toDate(input, "yyyy-MM-dd HH:mm:ss");
}
/**
 * 将字符串转成日期
 * @param input - 日期字符串
 * @param formatType - 格式化类型，如: yyyy-MM-dd HH:mm:ss
 * @return 日期类型，当出现异常时返回当前日期
 */
public static Date toDate(String input, String formatType){
    SimpleDateFormat format = new SimpleDateFormat(formatType);
    Date dt = new Date();
    if(DataValidator.isNullOrEmpty(input)){
        return dt;
    }
    try {
        dt = format.parse(input);
    } catch (ParseException e) {
    }
    return dt;
}
/**
 * 将字符串转成短日期格式 yyyy-MM-dd
 * @param input - 日期字符串
 * @return 日期类型，当出现异常时返回当前日期
 */
public static java.util.Date toShortDate(String input){
    return toDate(input, "yyyy-MM-dd");
}
/**
 * 将字符串转成长日期格式 yyyy-MM-dd HH:mm:ss
 * @param input - 日期字符串
```

第7章 综合实例——BLOG系统开发

```
    * @return 日期类型，当出现异常时返回当前日期
    */
    public static java.util.Date toFullDate(String input){
        return toDate(input, "yyyy-MM-dd HH:mm:ss");
    }
    /**
    * 将字符串转成整型
    * @param input - 要转换的字符串
    * @return 整数，出现异常则返回 0
    */
    public static int toInt(String input){
        try{
            return Integer.parseInt(input);
        }catch(Exception e){
            return 0;
        }
    }
}
```

3. DataValidator.java 实现

DataValidator.java 位于本工程的 com.ch7.common 目录，主要实现数据检验。DataValidator.java 实现代码如下：

```
package com.ch7.common;
import java.util.regex.Matcher;
import java.util.regex.Pattern;
/**
* 数据验证类
*/
public class DataValidator {
    /**
    * 验证字符串是否为空 = "" or = null
    * @param input - 需要验证的字符串
    * @return true/false
    */
    public static boolean isNullOrEmpty(String input){
        return "".equals(input) || input == null;
    }
    /**
    * 匹配正则表达式
    * @param input - 需要进行匹配的字符串
    * @param pattern - 正则表达式
    * @return true/false
    */
    public static boolean regexMatch(String input, String pattern)
    {
        if(isNullOrEmpty(input))
        {
            return false;
        }
```

```java
        return Pattern.compile(pattern).matcher(input).matches();
    }
    /**
     * TML 编码,支持换行符
     * @param input - 需要编码的字符串
     * @return 编码后的字符串
     */
    public static String htmlEncode(String input) {
        if(!DataValidator.isNullOrEmpty(input)) {
            input = input.replace("&", "&amp;");
            input = input.replace("<", "&lt;");
            input = input.replace(">", "&gt;");
            input = input.replace("'", "&#39;");
            input = input.replace("\"", "&quot;");
            input = input.replace("\r\n", "<br>");
            input = input.replace("\n", "<br>");
        }
        return input;
    }
    /**
     * HTML 反编码,支持换行符
     * @param input - 需要反编码的字符串
     * @return 编码前的原始字符串
     */
    public static String htmlDecode(String input){
        if(!DataValidator.isNullOrEmpty(input)) {
            input = input.replace("<br>", "\n");
            input = input.replace("&gt;", ">");
            input = input.replace("&lt;", "<");
            input = input.replace("&#39;", "'");
            input = input.replace("&quot;", "\"");
            input = input.replace("&amp;", "&");
        }
        return input;
    }
    /**
     * HTML 编码,不支持换行符
     * @param input - 需要编码的字符串
     * @return 编码后的字符串
     */
    public static String serverHtmlEncode(String input){
        if(DataValidator.isNullOrEmpty(input))
            return input;
        input = input.replace("&", "&amp;");
        input = input.replace("<", "&lt;");
        input = input.replace(">", "&gt;");
        input = input.replace("'", "&#39;");
        input = input.replace("\"", "&quot;");
        return input;
    }
```

```java
/**
 * HTML 反编码，不支持换行符
 * @param input - 需要反编码的字符串
 * @return 编码前的原始字符串
 */
public static String serverHtmlDecode(String input){
    if(DataValidator.isNullOrEmpty(input))
        return input;
    input = input.replace("&gt;", ">");
    input = input.replace("&lt;", "<");
    input = input.replace("&#39;", "'");
    input = input.replace("&quot;", "\"");
    input = input.replace("&amp;", "&");
    return input;
}
/**
 * 过滤所有 HTML 代码
 * @param input - 需要过滤的字符串
 * @return 过滤后的字符串
 */
public static String removeHtml(String input) {
    if(DataValidator.isNullOrEmpty(input))
        return input;
    Pattern p = Pattern.compile("<[^>]*>", Pattern.MULTILINE
            | Pattern.UNICODE_CASE);
    Matcher m = p.matcher(input);
    return m.replaceAll("");
}
}
```

4. MD5.java 实现

MD5.java 位于本工程的 com.ch7.common 目录，主要实现密码加密。MD5.java 实现代码如下：

```java
package com.ch7.common;
import java.io.UnsupportedEncodingException;
import java.security.MessageDigest;
import java.security.NoSuchAlgorithmException;
/**
 * MD5 加密类
 */
public class MD5 {
    private static MessageDigest digest = null;
    /**
     * 加密类，此方法默认为 16 位加密
     * @param data
     * @return
     */
    public synchronized static final String Encrypt(String data){
        return Encrypt(data, 16);
```

```java
}
/**
 * 加密类，此方法可以手动设置加密位数
 * @param data
 * @param len
 * @return
 */
public synchronized static final String Encrypt(String data, int len){
    if(digest == null) {
        try {
            digest = MessageDigest.getInstance("MD5");
        }
        catch (NoSuchAlgorithmException e) {
            e.printStackTrace();
        }
    }
    if(len != 16 && len != 32) len = 16;
    try {
        digest.update(data.getBytes("UTF-8"));
    } catch (UnsupportedEncodingException e) {
    }
    String s = encodeHex(digest.digest());
    if(len == 16){
      return s.substring(8, 24);
    }
    return s;
}
private static final String encodeHex(byte[] bytes) {
    int i;
    StringBuffer buf = new StringBuffer(bytes.length * 2);
    for(i = 0; i < bytes.length; i++) {
        if(((int) bytes[i] & 0xff) < 0x10) {
            buf.append("0");
        }
        buf.append(Long.toString((int) bytes[i] & 0xff, 16));
    }
    return buf.toString();
}
}
```

5. Utility.java 实现

Utility.java 位于本工程的 com.ch7.common 目录，主要实现一些常用通用方法。Utility.java 实现代码如下：

```java
package com.ch7.common;
import java.io.IOException;
import java.util.Calendar;
import java.util.Date;
import javax.servlet.ServletException;
import javax.servlet.http.Cookie;
```

```java
import javax.servlet.http.HttpServletRequest;
import javax.servlet.http.HttpServletResponse;
import javax.servlet.jsp.PageContext;
/**
 * 基类
 */
public class Utility {
    /**
     * 截取字符串
     *
     * @param input
     * @param len
     * @return
     */
    public static String Substring(String input, int len) {
        if(DataValidator.isNullOrEmpty(input))
            return "";
        if(len >= input.length())
            return input;
        return input.substring(0, len);
    }
    /**
     * 获取当前时间
     *
     * @return
     */
    public static Date getNowDateTime() {
        Calendar cal = Calendar.getInstance();
        String now = cal.get(Calendar.YEAR) + "-"
                + (cal.get(Calendar.MONTH) + 1) + "-" + cal.get(Calendar.DATE)
                + " " + cal.get(Calendar.HOUR) + ":" + cal.get(Calendar.MINUTE)
                + ":" + cal.get(Calendar.SECOND);
        return DataConverter.toDate(now, "yyyy-MM-dd HH:mm:ss");
    }
    /**
     * 获得当前日期
     * @return
     */
    public static Date getNowDate() {
        Calendar cal = Calendar.getInstance();
        String now = cal.get(Calendar.YEAR) + "-"
                + (cal.get(Calendar.MONTH) + 1) + "-" + cal.get(Calendar.DATE);
        return DataConverter.toDate(now, "yyyy-MM-dd");
    }
    /**
     * 写入cookie
     *
```

```java
     * @param response
     * @param key
     * @param value
     */
    public static void writeCookie(HttpServletResponse response, String key,
            String value) {
        writeCookie(response, key, value, -1);
    }
    /**
     * 写入cookie
     *
     * @param response
     * @param key
     * @param value
     * @param expirse
     */
    public static void writeCookie(HttpServletResponse response, String key,
            String value, int expirse) {
        Cookie newCookie = new Cookie(key, value);
        if(expirse > 0)
           expirse = expirse * 60;
        newCookie.setPath("/");
        newCookie.setMaxAge(expirse);
        response.addCookie(newCookie);
    }
    /**
     * 读取cookie值
     *
     * @param request
     * @param key
     * @return
     */
    public static String readCookie(HttpServletRequest request, String key) {
        String value = "";
        Cookie[] ck = request.getCookies();
        if(ck == null)
           return "";
        for(Cookie c : ck) {
           if(c.getName().equals(key)) {
               value = c.getValue();
               break;
           }
        }
        return value;
    }
}
```

7.4.2 数据访问层功能实现

1. Blog.java 实现

Blog.java 位于本工程的 com.ch7.dal 目录。Blog.java 实现代码如下:

```java
package com.ch7.dal;

import java.sql.ResultSet;
import java.sql.SQLException;
import java.util.ArrayList;
import java.util.List;
import com.ch7.common.Conn;
import com.ch7.common.DataValidator;
import com.ch7.model.BlogInfo;
public class Blog {
    Conn conn = new Conn();
    /**
     * 获取博文列表
     *
     * @return
     * @throws SQLException
     */
    public List<BlogInfo> getList(String keyword) throws SQLException {
        List<BlogInfo> list = new ArrayList<BlogInfo>();
        String sql = "select b.*,c.name as ClassName from Blog b left join class c on b.classid=c.id ";
        if(DataValidator.isNullOrEmpty(keyword)) {
            sql = sql + " order by id desc";
        } else {
            sql = sql + " where b.title like '%" + keyword
                + "%' order by id desc";
        }
        ResultSet rs = conn.executeQuery(sql);
        while(rs.next()) {
            BlogInfo info = new BlogInfo();
            info.setId(rs.getInt("Id"));
            info.setTitle(rs.getString("Title"));
            info.setContext(rs.getString("Context"));
            info.setCreatedtime(rs.getDate("CreatedTime"));
            info.setClassid(rs.getInt("ClassId"));
            info.setClassName(rs.getString("ClassName"));
            list.add(info);
        }
        conn.close();
        return list;
    }
    /**
     * 获得某分类下面的所有博文列表
     *
     * @param classId
```

```java
     * @return
     * @throws SQLException
     */
    public List<BlogInfo> getListByClassId(int classId) throws SQLException {
        List<BlogInfo> list = new ArrayList<BlogInfo>();
        String sql = "select b.*,c.name as ClassName from Blog b left join class c on b.classid=c.id where b.classId="
            + classId + " order by id desc";
        ResultSet rs = conn.executeQuery(sql);
        while(rs.next()) {
            BlogInfo info = new BlogInfo();
            info.setId(rs.getInt("Id"));
            info.setTitle(rs.getString("Title"));
            info.setContext(rs.getString("Context"));
            info.setCreatedtime(rs.getDate("CreatedTime"));
            info.setClassid(rs.getInt("ClassId"));
            info.setClassName(rs.getString("ClassName"));
            list.add(info);
        }
        conn.close();
        return list;
    }
    /**
     * 获得单条博文
     *
     * @param id
     * @return
     * @throws SQLException
     */
    public BlogInfo getBlogInfo(int id) throws SQLException {
        BlogInfo info = new BlogInfo();
        String sql = "select b.*,c.name as ClassName from Blog b left join class c on b.classid=c.id where b.id="
            + id + "";
        ResultSet rs = conn.executeQuery(sql);
        if(rs.next()) {
            info.setId(rs.getInt("Id"));
            info.setTitle(rs.getString("Title"));
            info.setContext(rs.getString("Context"));
            info.setCreatedtime(rs.getDate("CreatedTime"));
            info.setClassid(rs.getInt("ClassId"));
            info.setClassName(rs.getString("ClassName"));
        }
        conn.close();
        return info;
    }
    /**
     * 博文插入操作
     *
```

```java
 * @param info
 * @return
 */
public int insert(BlogInfo info) {
    String sql = "insert into Blog(Title,Context,CreatedTime,ClassId) values ";
    sql = sql + " ('" + info.getTitle() + "','" + info.getContext()
            + "',now()," + info.getClassid() + ")";
    int result = 0;
    System.out.print(sql);
    result = conn.executeUpdate(sql);
    conn.close();
    return result;
}
/**
 * 博文修改
 *
 * @param info
 * @return
 */
public int update(BlogInfo info) {
    String sql = "update Blog set" + " Title='" + info.getTitle()
            + "',Context='" + info.getContext() + "',ClassId='"
            + info.getClassid() + "' where id=" + info.getId() + "";
    int result = 0;
    System.out.print(sql);
    result = conn.executeUpdate(sql);
    conn.close();
    return result;
}
/**
 * 博文删除
 *
 * @param id
 * @return
 */
public int delete(int id) {
    String sql = "delete from Blog where id =" + id + "";
    int result = 0;
    result = conn.executeUpdate(sql);
    conn.close();
    return result;
}
```

2. Class.java 实现

Class.java 位于本工程的 com.ch7.dal 目录。Class.java 实现代码如下：

```java
package com.ch7.dal;

import java.sql.ResultSet;
```

```java
import java.sql.SQLException;
import java.util.ArrayList;
import java.util.List;
import com.ch7.common.Conn;
import com.ch7.model.ClassInfo;

public class Class {
    Conn conn = new Conn();
    /**
     * 获取博文分类列表
     *
     * @return
     * @throws SQLException
     */
    public List<ClassInfo> getList() throws SQLException {
        List<ClassInfo> list = new ArrayList<ClassInfo>();
        String sql = "select * from Class order by Sort asc";
        ResultSet rs = conn.executeQuery(sql);
        while(rs.next()) {
            ClassInfo info = new ClassInfo();
            info.setId(rs.getInt("Id"));
            info.setName(rs.getString("Name"));
            info.setSort(rs.getInt("Sort"));
            list.add(info);
        }
        conn.close();
        return list;
    }
    /**
     * 获得单条分类信息
     *
     * @param id
     * @return
     * @throws SQLException
     */
    public ClassInfo getClassInfo(int id) throws SQLException {
        ClassInfo info = new ClassInfo();
        String sql = "select * from Class c where id=" + id + "";
        ResultSet rs = conn.executeQuery(sql);
        if(rs.next()) {
            info.setId(rs.getInt("Id"));
            info.setName(rs.getString("Name"));
            info.setSort(rs.getInt("Sort"));
        }
        conn.close();
        return info;
    }

    /**
     * 博文分类插入
```

```java
     *
     * @param info
     * @return
     */
    public int insert(ClassInfo info) {
        String sql = "insert into Class(Name,Sort) values ";
        sql = sql + " ('" + info.getName() + "','" + info.getSort() + "')";
        int result = 0;
        result = conn.executeUpdate(sql);
        conn.close();
        return result;
    }
    /**
     * 博文分类修改
     *
     * @param info
     * @return
     */
    public int update(ClassInfo info) {
        String sql = "update Class set" + " Name='" + info.getName()
                + "',Sort=" + info.getSort() + " where id=" + info.getId() + "";
        int result = 0;
        result = conn.executeUpdate(sql);
        conn.close();
        return result;
    }
    /**
     * 博文分类删除
     *
     * @param id
     * @return
     */
    public int delete(int id) {
        String sql = "delete from Class where id =" + id + "";
        int result = 0;
        result = conn.executeUpdate(sql);
        conn.close();
        return result;
    }
}
```

3. Comment.java 实现

Comment.java 位于本工程的 com.ch7.dal 目录。Comment.java 实现代码如下：

```java
package com.ch7.dal;
import java.sql.ResultSet;
import java.sql.SQLException;
import java.util.ArrayList;
import java.util.List;
import com.ch7.common.Conn;
import com.ch7.model.CommentInfo;
```

```java
public class Comment {
    Conn conn = new Conn();
    /**
     * 获取博文评论列表
     *
     * @return
     * @throws SQLException
     */
    public List<CommentInfo> getList() throws SQLException {
        List<CommentInfo> list = new ArrayList<CommentInfo>();
        String sql = "select * from Comment order by id desc";
        ResultSet rs = conn.executeQuery(sql);

        while (rs.next()) {
            CommentInfo info = new CommentInfo();
            info.setId(rs.getInt("Id"));
            info.setContext(rs.getString("Context"));
            info.setBlogid(rs.getInt("BlogId"));
            info.setCreatedtime(rs.getDate("CreatedTime"));
            info.setUsername(rs.getString("UserName"));
            list.add(info);
        }
        conn.close();
        return list;
    }
    /**
     * 获得单条评论信息
     *
     * @param id
     * @return
     * @throws SQLException
     */
    public CommentInfo getCommentInfo(int id) throws SQLException {
        CommentInfo info = new CommentInfo();
        String sql = "select * from Comment c where id=" + id + "";
        ResultSet rs = conn.executeQuery(sql);
        if (rs.next()) {
            info.setId(rs.getInt("Id"));
            info.setContext(rs.getString("Context"));
            info.setBlogid(rs.getInt("BlogId"));
            info.setCreatedtime(rs.getDate("CreatedTime"));
            info.setUsername(rs.getString("UserName"));
        }
        conn.close();
        return info;
    }
    /**
     * 获得某博文的所有评论
     *
     * @param id
```

第 7 章 综合实例——BLOG 系统开发

```java
     * @return
     * @throws SQLException
     */
    public List<CommentInfo> getListByBlogId(int blogid) throws
SQLException {
        List<CommentInfo> list = new ArrayList<CommentInfo>();
        String sql = "select * from Comment where blogid=" + blogid
            + " order by id desc";
        ResultSet rs = conn.executeQuery(sql);
        while (rs.next()) {
            CommentInfo info = new CommentInfo();
            info.setId(rs.getInt("Id"));
            info.setContext(rs.getString("Context"));
            info.setBlogid(rs.getInt("BlogId"));
            info.setCreatedtime(rs.getDate("CreatedTime"));
            info.setUsername(rs.getString("UserName"));
            list.add(info);
        }
        conn.close();
        return list;
    }
    /**
     * 博文评论插入
     *
     * @param info
     * @return
     */
    public int insert(CommentInfo info) {
        String sql = "insert into Comment(Context,BlogId,CreatedTime,
UserName) values ";
        sql = sql + " ('" + info.getContext() + "'," + info.getBlogid()
            + ",now(),'" + info.getUsername() + "')";
        int result = 0;
        System.out.println(sql);
        result = conn.executeUpdate(sql);
        conn.close();
        return result;
    }
    /**
     * 博文评论修改
     *
     * @param info
     * @return
     */
    public int update(CommentInfo info) {
        String sql = "update Comment set" + " Context='" + info.getContext()
            + "',BlogId=" + info.getBlogid() + ",CreatedTime='"
            + info.getCreatedtime() + "',UserName = '" + info.getUsername()
            + "' where id=" + info.getId() + "";
        int result = 0;
```

```java
        result = conn.executeUpdate(sql);
        conn.close();
        return result;
    }
    /**
     * 博文评论删除
     * @param id
     * @return
     */
    public int delete(int id) {
        String sql = "delete from Comment where id =" + id + "";
        int result = 0;
        result = conn.executeUpdate(sql);
        conn.close();
        return result;
    }
}
```

4. Users.java 实现

Users.java 位于本工程的 com.ch7.dal 目录。Users.java 实现代码如下:

```java
package com.ch7.dal;
import java.sql.ResultSet;
import java.sql.SQLException;
import java.util.ArrayList;
import java.util.List;
import com.ch7.common.Conn;
import com.ch7.common.MD5;
import com.ch7.model.UsersInfo;

public class Users {
    Conn conn = new Conn();
    /**
     * 获取用户列表
     *
     * @return
     * @throws SQLException
     */
    public List<UsersInfo> getList() throws SQLException {
        List<UsersInfo> list = new ArrayList<UsersInfo>();
        String sql = "select * from Users order by username asc";
        ResultSet rs = conn.executeQuery(sql);
        while (rs.next()) {
            UsersInfo info = new UsersInfo();
            info.setUsername(rs.getString("UserName"));
            info.setPassword(rs.getString("Password"));
            info.setEmail(rs.getString("Email"));
            info.setPower(rs.getString("Power"));
            list.add(info);
        }
        conn.close();
```

第 7 章 综合实例——BLOG 系统开发

```java
        return list;
    }
    /**
     * 判断当前登录用户是否存在
     *
     * @param id
     * @return
     * @throws SQLException
     */
    public boolean isExist(String username, String password)
            throws SQLException {
        boolean result = false;
        UsersInfo info = new UsersInfo();
        String sql = "select * from Users u where UserName='" + username
                + "' and Password='" + password + "'";
        System.out.println(sql);
        ResultSet rs = conn.executeQuery(sql);
        if(rs.next()) {
            info.setUsername(rs.getString("UserName"));
            info.setPassword(rs.getString("Password"));
            info.setEmail(rs.getString("Email"));
            info.setPower(rs.getString("Power"));
            result = true;
        }
        conn.close();
        return result;
    }
    /**
     * 获得单个用户信息
     *
     * @param id
     * @return
     * @throws SQLException
     */
    public UsersInfo getUsersInfo(String username) throws SQLException {
        UsersInfo info = new UsersInfo();
        String sql = "select * from Users U where UserName='" + username + "'";
        ResultSet rs = conn.executeQuery(sql);
        if(rs.next()) {
            info.setUsername(rs.getString("UserName"));
            info.setPassword(rs.getString("Password"));
            info.setEmail(rs.getString("Email"));
            info.setPower(rs.getString("Power"));
        }
        conn.close();
        return info;
    }
    /**
     * 判断注册用户是否已存在
     *
```

```java
 * @param id
 * @return
 * @throws SQLException
 */
public boolean isExistUsersInfo(String username) throws SQLException {
    boolean result = false;
    UsersInfo info = new UsersInfo();
    String sql = "select * from Users U where UserName='" + username + "'";
    ResultSet rs = conn.executeQuery(sql);
    if(rs.next()) {
        info.setUsername(rs.getString("UserName"));
        info.setPassword(rs.getString("Password"));
        info.setEmail(rs.getString("Email"));
        info.setPower(rs.getString("Power"));
        result = true;
    }
    conn.close();
    return result;
}
/**
 * 用户插入
 *
 * @param info
 * @return
 */
public int insert(UsersInfo info) {
    String sql = "insert into Users(UserName,Password,Email,Power) values ";
    sql = sql + " ('" + info.getUsername() + "','" + info.getPassword()
            + "','" + info.getEmail() + "','" + info.getPower() + "')";
    int result = 0;
    result = conn.executeUpdate(sql);
    conn.close();
    return result;
}
/**
 * 用户修改
 *
 * @param info
 * @return
 */
public int update(UsersInfo info) {
    String sql = "update Users set Password='"
            + MD5.Encrypt(info.getPassword()) + "',Email='"
            + info.getEmail() + "',Power = '" + info.getPower()
            + "' where UserName='" + info.getUsername() + "'";
    int result = 0;
    result = conn.executeUpdate(sql);
    conn.close();
    return result;
}
```

```java
/**
 * 用户删除
 *
 * @param id
 * @return
 */
public int delete(String username) {
    String sql = "delete from Users where UserName ='" + username + "'";
    int result = 0;
    result = conn.executeUpdate(sql);
    conn.close();
    return result;
}
```

7.4.3 后台表示层功能实现

1. 验证后台管理员是否登录本系统 islogin.jsp

islogin.jsp 位于 WebRoot\manager 目录，核心代码如下：

```jsp
<%@ page language="java" import="java.util.*" pageEncoding="utf-8"%><%@page
    import="com.ch7.common.Utility"%><%@page
    import="com.ch7.common.DataValidator"%>
<%
    String data = Utility.readCookie(request, "admin");
    if(DataValidator.isNullOrEmpty(data)) {
        response.sendRedirect("login.jsp");
    }
%>
```

2. 后台登录页 login.jsp

login.jsp 位于 WebRoot\manager 目录，核心代码如下：

```html
<script language="javascript" type="text/javascript">
function check(form){
    if(document.loginform.AdminName.value == ""){
        alert("请输入管理员名");
        document.loginform.AdminName.focus();
        return false;
    }
    if(document.loginform.Password.value == ""){
        alert("请输入登录密码");
        document.loginform.Password.focus();
        return false;
    }
}
window.onload = function(){
    document.getElementById("AdminName").focus();
    document.getElementById("AdminName").value = "";
    document.getElementById("Password").value = "";
```

```
        }
    </script>
    ...
    <form name="loginform" action="./login-check.jsp" method="post" onsubmit="return check(this)">
        <table width="350" border="0" cellspacing="1" cellpadding="0">
            <tr>
                <td colspan="2" align="center"><strong>博客系统管理员登录</strong></td>
            </tr>
            <tr>
                <td width="101" class="item">管理员名: </td>
                <td width="246" class="input"><input name="AdminName" type="text" id="AdminName" size="30" /></td>
            </tr>
            <tr>
                <td class="item">登录密码: </td>
                <td class="input"><input name="Password" type="password" id="Password" size="30" /></td>
            </tr>
            <tr>
                <td colspan="2" class="button"><input type="submit" value="登录 " /></td>
            </tr>
        </table>
    </form>
    ...
```

3. 后台登录数据处理页 login-check.jsp

login-check.jsp 位于 WebRoot\manager 目录，核心代码如下：

```
<%@ page language="java" import="java.util.*" pageEncoding="utf-8"%>
<%@page import="com.ch7.dal.Users"%>
<%@page import="com.ch7.model.UsersInfo"%>
<%@page import="com.ch7.common.MD5"%>
<%@page import="com.ch7.common.DataValidator"%><%@page
    import="com.ch7.common.Utility"%>
<%
    Users users = new Users();
    String username = request.getParameter("AdminName");
    String password = MD5.Encrypt(request.getParameter("Password"));
    if(!users.isExist(username, password)) {
        out.println(" <script>alert('用户名密码有误');window.location.href('login.jsp');</script>");
    } else {
        Utility.writeCookie(response, "admin", username);
        response.sendRedirect("index.htm");
    }
%>
```

4. 后台框架页 index.htm

index.htm 位于 WebRoot\manager 目录，核心代码如下：

```html
<!DOCTYPE html PUBLIC "-//W3C//DTD XHTML 1.0 Frameset//EN" "http://www.w3.org/TR/xhtml1/DTD/xhtml1-frameset.dtd">
<html xmlns="http://www.w3.org/1999/xhtml">
<head>
<meta http-equiv="Content-Type" content="text/html; charset=utf-8" />
<title>博客管理系统后台</title>
</head>
<frameset rows="*" cols="*,900" framespacing="0" frameborder="no" border="0">
  <frame src="menu.jsp" name="leftFrame" id="leftFrame" title="mainFrame" />
  <!--rigtht.jsp 为一个空白页-->
  <frame src="right.jsp" name="rightFrame" scrolling="no" noresize="noresize" id="rightFrame" title="rightFrame" />
</frameset>
<noframes><body>
</body></noframes>
</html>
```

5. 后台导航页 menu.jsp

menu.jsp 位于 WebRoot\manager 目录，核心代码如下：

```html
<table width="256" border="0">
          <tr><td>博客管理系统后台</td>
          </tr>
          <tr><td>欢迎您：<%=Utility.readCookie(request, "admin")%><a href="logout.jsp" target="_parent">退出</a></td>
          </tr>
          <tr><td><a href="/blog/manager/blog-edit.jsp?action=add" target="rightFrame">发布博文</a></td>
          </tr>
          <tr><td><a href="/blog/manager/blog-manage.jsp" target="rightFrame">博文管理</a></td>
          </tr>
          <tr><td> </td>
          </tr>
          <tr><td><a href="/blog/manager/class-edit.jsp?action=add" target="rightFrame">添加博文分类</a></td>
          </tr>
          <tr><td><a href="/blog/manager/class-manage.jsp" target="rightFrame">博文分类管理</a></td>
          </tr>
          <tr><td> </td>
          </tr>
          <tr><td><a href="/blog/manager/users-manage.jsp" target="rightFrame">用户管理</a></td>
          </tr>
      </table>
```

6. 后台博文管理页 blog-manage.jsp

blog-manage.jsp 位于 WebRoot\manager 目录，核心代码如下：

```jsp
<%@ page language="java" import="java.util.*" pageEncoding="utf-8"%>
<%@include file="islogin.jsp"%>
<%@page import="com.ch7.model.BlogInfo"%>
<%@page import="com.ch7.dal.Blog"%>
<%
    request.setCharacterEncoding("utf-8");
    Blog blog = new Blog();
    String keyword = request.getParameter("keyword");
    List<BlogInfo> list = blog.getList(keyword);
%>
   ...
<body>
        <p>当前位置：博文管理</p>
        <form id="form1" name="form1" method="post" action="blog-manage.jsp">
            查询条件：博文标题
            <input type="text" name="keyword" id="keyword" />
            <input type="submit" name="button" id="button" value="查询" />
        </form>
        <table width="98%" border="1">
            <tr>
                <td>博文标题</td>
                <td>所属分类</td>
                <td>发布时间</td>
                <td>操作</td>
            </tr>
            <%
            for(BlogInfo info : list) { //遍历输出list集合中的数据
            %>
            <tr>
                <td><%=info.getTitle()%></td>
                <td><%=info.getClassName()%></td>
                <td><%=info.getCreatedtime()%></td>
                <td><a href="blog-edit.jsp?id=<%=info.getId()%>&action=edit">编辑</a> |
                    <a href="blog-delete.jsp?id=<%=info.getId()%>">删除</a> | <a href="comment-manage.jsp?blogid=<%=info.getId()%>">评论</a>
                </td>
            </tr>
            <%
            }
            %>
        </table>
</body>
```

7. 后台博文编辑与添加页 blog-edit.jsp

blog-edit.jsp 位于 WebRoot\manager 目录，核心代码如下：

```jsp
<%@ page language="java" import="java.util.*,net.fckeditor.*"
```

第7章 综合实例——BLOG 系统开发

```jsp
        pageEncoding="utf-8"%>
<%@include file="islogin.jsp"%>
<%@page import="com.ch7.dal.Class"%>
<%@page import="com.ch7.model.ClassInfo"%>
<%@page import="com.ch7.common.DataConverter"%>
<%@page import="com.ch7.model.BlogInfo"%>
<%@page import="com.ch7.dal.Blog"%>
<%@page import="com.ch7.common.*"%>
<%
    request.setCharacterEncoding("utf-8");
    String action = request.getParameter("action");
    String pageAction = "";
    int id = DataConverter.toInt(request.getParameter("id"));
    BlogInfo info = new BlogInfo();
    Blog blog = new Blog();
    //保存
    if("update".equals(action) || "insert".equals(action)) {
        if("update".equals(action)) {
            info = blog.getBlogInfo(id);
            if(info == null) {
                out.println("<script>alert('博文ID有误');window.location.href('login.jsp');</script>");
            }
            info.setId(id);
        }
        info.setTitle(request.getParameter("txtTitle"));
        info.setClassid(DataConverter.toInt(request
            .getParameter("selClass")));
        info.setContext(request.getParameter("content"));
        if("insert".equals(action)) {
            info.setCreatedtime(Utility.getNowDateTime());
            blog.insert(info);
        } else
            blog.update(info);
        response.sendRedirect("blog-manage.jsp");
    }
    //编辑
    if("edit".equals(action) || "add".equals(action)) {
        info = blog.getBlogInfo(id);
        if(info == null) {
            // Utility.showErrorMessage(pageContext, "学生ID错误");
        }
        pageAction = "update";
    }
    // 添加
    if("add".equals(action)) {
        pageAction = "insert";
    }
    //调用在线编辑器
    FCKeditor fckEditor = new FCKeditor(request, "content");
```

```jsp
        fckEditor.setHeight("400");
        fckEditor.setValue(info.getContext());
        //获得博文分类
        Class cls = new Class();
        List<ClassInfo> list = cls.getList();
%>
…
<script type="text/javascript">
function FCKeditor_OnComplete(editorInstance) {
    window.status = editorInstance.Description;
}
</script>
…
<body>
        <p>当前位置：博文编辑/添加</p>
        <form id="form1" name="form1" method="post" action="blog-edit.jsp">
            <table width="100%" border="1">
                <tr>
                    <td>博文标题</td>
                    <td>
                        <input type="text" name="txtTitle" id="txtTitle" width="500px"
                            value="<%=info.getTitle()%>" />
                    </td>
                </tr>
                <tr>
                    <td>博文所属分类</td>
                    <td>
                        <select name="selClass" id="selClass">
                            <%
                                for(ClassInfo cinfo : list) {
                            %>
                            <option value="<%=cinfo.getId()%>"
                                <%if(cinfo.getId() == info.getClassid())
                            out.print("selected");%>>
                                <%=cinfo.getName()%></option>
                            <%
                                }
                            %>
                        </select>
                    </td>
                </tr>
                <tr>
                    <td>博文内容</td>
                    <td><%=fckEditor%></td>
                </tr>
                <tr>
                    <td colspan="2">
<input type="submit" name="button" id="button" value="提交" />
<input type="reset" name="button2" id="button2" value="重置" />
```

第 7 章 综合实例——BLOG 系统开发

```
            <input type="hidden" name="action" value="<%=pageAction%>" />
            <input type="hidden" name="id" value="<%=info.getId()%>" />
                    </td>
                </tr>
            </table>
        </form>
    </body>
```

8. 后台博文删除页 blog-delete.jsp

blog-delete.jsp 位于 WebRoot\manager 目录，核心代码如下：

```jsp
<%@ page language="java" import="java.util.*" pageEncoding="utf-8"%>
<%@include file="islogin.jsp"%>
<%@page import="com.ch7.dal.Blog"%><%@page
    import="com.ch7.common.DataConverter"%>
<%
    Blog blog = new Blog(); //创建对象
    int result = 0;
    result = blog
            .delete(DataConverter.toInt(request.getParameter("id")));
    if(result == 1) {
        out.println("<script>alert('博文删除成功');window.location.href('blog-manage.jsp');</script>");
    } else {
        out.println("<script>alert('博文删除失败');window.location.href('blog-manage.jsp');</script>");
    }
%>
```

9. 后台博文评论页 comment-manage.jsp

comment-manage.jsp 位于 WebRoot\manager 目录，核心代码如下：

```jsp
<%@ page language="java" import="java.util.*"
pageEncoding="utf-8"%><%@page
    import="com.ch7.dal.Comment"%><%@page
    import="com.ch7.model.CommentInfo"%><%@page
    import="com.ch7.common.DataConverter"%>
<%@include file="islogin.jsp"%>
<%
    request.setCharacterEncoding("utf-8");
    Comment comment = new Comment();
    String keyword = request.getParameter("keyword");
    int blogid = DataConverter.toInt(request.getParameter("blogid"));
    List<CommentInfo> list = comment.getListByBlogId(blogid);
%>
…
<body>
        <p>
            当前位置：评论管理
        </p>
        <table width="98%" border="1">
```

```jsp
            <tr>
                <td>评论内容 </td>
                <td>评论人</td>
                <td>评论时间</td>
                <td>操作</td>
            </tr>
            <%
                for(CommentInfo info : list) { //遍历输出list集合中的数据
            %>
            <tr>
                <td><%=Utility.Substring(
                    DataValidator.removeHtml(info.getContext()), 100)%></td>
                <td><%=info.getUsername()%></td>
                <td><%=info.getCreatedtime()%></td>
                <td>
                    <a href="comment-delete.jsp?id=<%=info.getId()%>&blogid=<%=info.getBlogid()%>">删除</a>
                </td>
            </tr>
            <%
                }
            %>
        </table>
    </body>
```

10. 后台博文评论删除页 comment-delete.jsp

Comment-delete.jsp 位于 WebRoot\manager 目录，核心代码如下：

```jsp
<%@ page language="java" import="java.util.*" pageEncoding="utf-8"%>
<%@include file="islogin.jsp"%>
<%@page import="com.ch7.dal.Comment"%>
<%@page import="com.ch7.common.DataConverter"%>
<%
    Comment comment = new Comment(); //创建对象
    int result = 0;
    result = comment
            .delete(DataConverter.toInt(request.getParameter("id")));
    if(result == 1) {
        out.println("<script>alert('博文评论删除成功');window.location.href('comment-manage.jsp?blogid="+request.getParameter("blogid")+"');</script>");
    } else {
        out.println("<script>alert('博文评论删除失败');window.location.href('comment-manage.jsp?blogid="+request.getParameter("blogid")+"');</script>");
    }
%>
```

11. 后台博文分类管理页 class-manage.jsp

class-manage.jsp 位于 WebRoot\manager 目录，核心代码如下：

```jsp
<%@ page language="java" import="java.util.*" pageEncoding="utf-8"%>
```

```jsp
<%@include file="islogin.jsp"%>
<%@page import="com.ch7.dal.Class"%>
<%@page import="com.ch7.model.ClassInfo"%>
<%
    request.setCharacterEncoding("utf-8");
    Class cls = new Class();
    String keyword = request.getParameter("keyword");
    List<ClassInfo> list = cls.getList();
%>
...
<body>
        <p>
                当前位置：分类管理
        </p>
        <table width="98%" border="1">
            <tr>
                <td>名称</td>
                <td>排序</td>
                <td>操作</td>
            </tr>
            <%
                for (ClassInfo info : list) { //遍历输出list集合中的数据
            %>
            <tr>
                <td><%=info.getName()%></td>
                <td><%=info.getSort()%></td>
                <td>
                    <a href="class-edit.jsp?id=<%=info.getId()%>&action=edit">编辑</a> |
                    <a href="class-delete.jsp?id=<%=info.getId()%>">删除</a>
                </td>
            </tr>
            <%
                }
            %>
        </table>
</body>
```

12. 后台博文分类编辑\添加页 class-edit.jsp

class-edit.jsp 位于 WebRoot\manager 目录，核心代码如下：

```jsp
<%@ page language="java" import="java.util.*,net.fckeditor.*"
    pageEncoding="utf-8"%>
<%@include file="islogin.jsp"%>
<%@page import="com.ch7.common.DataConverter"%>
<%@page    import="com.ch7.model.ClassInfo"%>
<%@page import="com.ch7.dal.Class"%>
<%
    request.setCharacterEncoding("utf-8");
    String action = request.getParameter("action");
    String pageAction = "";
```

```jsp
        int id = DataConverter.toInt(request.getParameter("id"));
        ClassInfo info = new ClassInfo();
        Class cls = new Class();
        //保存
        if("update".equals(action) || "insert".equals(action)) {
            if("update".equals(action)) {
                info = cls.getClassInfo(id);
                if(info == null) {
                    out.println("<script>alert('博文ID有误');window.location.href('login.jsp');</script>");
                }
                info.setId(id);
            }
            info.setName(request.getParameter("txtName"));
            info.setSort(DataConverter.toInt(request
                .getParameter("txtSort")));
            if("insert".equals(action)) {
                cls.insert(info);
            } else
                cls.update(info);
            response.sendRedirect("class-manage.jsp");
        }
        //编辑
        if("edit".equals(action) || "add".equals(action)) {
            info = cls.getClassInfo(id);
            if(info == null) {
                // Utility.showErrorMessage(pageContext, "学生ID错误");
            }
            pageAction = "update";
        }
        // 添加
        if("add".equals(action)) {
            pageAction = "insert";
        }
%>
...
<body>
    <p>
        当前位置：分类编辑/添加
    </p>
    <form id="form1" name="form1" method="post" action="class-edit.jsp">
        <table width="100%" border="1">
            <tr>
                <td>分类名称</td>
                <td>
                    <input type="text" name="txtName" id="" txtName"" width="500px"
                        value="<%=info.getName()%>" />
                </td>
            </tr>
```

```
            <tr>
                <td>排序</td>
                <td>
                    <input type="text" name="txtSort" id="" txtSort"" width="500px"
                        value="<%=info.getSort()%>" />
                </td>
            </tr>
            <tr>
                <td colspan="2">
    <input type="submit" name="button" id="button" value="提交" />
    <input type="reset" name="button2" id="button2" value="重置" />
    <input type="hidden" name="action" value="<%=pageAction%>" />
    <input type="hidden" name="id" value="<%=info.getId()%>" />
                </td>
            </tr>
        </table>
    </form>
</body>
```

13. 后台博文分类删除页 class-delete.jsp

class-delete.jsp 位于 WebRoot\manager 目录，核心代码如下：

```
<%@ page language="java" import="java.util.*" pageEncoding="utf-8"%>
<%@include file="islogin.jsp"%>
<%@page import="com.ch7.common.DataConverter"%>
<%@page    import="com.ch7.dal.Class"%>
<%
    Class cls = new Class(); //创建对象
    int result = 0;
    result = cls
            .delete(DataConverter.toInt(request.getParameter("id")));
    if(result == 1) {
        out.println("<script>alert('分类删除成功');window.location.href('class-manage.jsp');</script>");
    } else {
        out.println("<script>alert('分类删除失败');window.location.href('class-manage.jsp');</script>");
    }
%>
```

14. 后台用户管理页 users-manage.jsp

users-manage.jsp 位于 WebRoot\manager 目录，核心代码如下：

```
<%@ page language="java" import="java.util.*" pageEncoding="utf-8"%>
<%@include file="islogin.jsp"%>
<%@page import="com.ch7.model.UsersInfo"%>
<%@page    import="com.ch7.dal.Users"%>
<%
    request.setCharacterEncoding("utf-8");
    Users user = new Users();
    String keyword = request.getParameter("keyword");
```

```jsp
        List<UsersInfo> list = user.getList();
%>
...
<body>
    <p>
        当前位置：用户管理
    </p>
    <table width="98%" border="1">
        <tr>
            <td>用户名</td>
            <td>Email</td>
            <td>身份</td>
            <td>操作</td>
        </tr>
        <%
        for(UsersInfo info : list) { //遍历输出list集合中的数据
        %>
        <tr>
            <td><%=info.getUsername()%></td>
            <td><%=info.getEmail()%></td>
            <td><%=info.getPower()%></td>
            <td>
                <a
                    href="users-edit.jsp?username=<%=info.getUsername()%>&action=edit">编辑</a>
                |
                <a href="users-delete.jsp?username=<%=info.getUsername()%>">删除</a>
            </td>
        </tr>
        <%
        }
        %>
    </table>
</body>
```

15. 后台用户编辑\添加页 users-edit.jsp

users-edit.jsp 位于 WebRoot\manager 目录，核心代码如下：

```jsp
<%@ page language="java" import="java.util.*,net.fckeditor.*"
    pageEncoding="utf-8"%>
<%@include file="islogin.jsp"%>
<%@page import="com.ch7.common.DataConverter"%>
<%@page import="com.ch7.common.*"%>
<%@page    import="com.ch7.model.UsersInfo"%>
<%@page import="com.ch7.dal.Users"%>
<%
    request.setCharacterEncoding("utf-8");
    String action = request.getParameter("action");
    String pageAction = "";
    String username = request.getParameter("username");
```

```
            UsersInfo info = new UsersInfo();
            Users user = new Users();
            //保存
            if("update".equals(action)) {
                info = user.getUsersInfo(username);
                if(info == null) {
                    out.println("<script>alert('用户ID有误');window.location.href
('users-manage.jsp');</script>");
                }
                info.setUsername(username);
                info.setPassword(request.getParameter("txtPassword"));
                String email = "";
                if(DataValidator.isEmail(request.getParameter("txtEmail"))) {
                    email = request.getParameter("txtEmail");
                } else {
                    email = "";
                }
                info.setEmail(email);
                info.setPower(request.getParameter("selPower"));
                if ("update".equals(action)) {
                    user.update(info);
                }
                response.sendRedirect("users-manage.jsp");
            }
            //编辑
            if("edit".equals(action)) {
                info = user.getUsersInfo(username);
                if(info == null) {
                    // Utility.showErrorMessage(pageContext, "学生ID错误");
                }
                pageAction = "update";
            }
        %>
        ...
        <body>
            <p>
                当前位置：用户编辑/添加
            </p>
            <form id="form1" name="form1" method="post" action="users-edit.jsp">
                <table width="100%" border="1">
                    <tr>
                        <td>用户名</td>
                        <td>
                            <input type="text" name="txtUserName" id="txtUserName"
                                width="500px" value="<%=info.getUsername()%>"
                                readonly="readonly" />
                        </td>
                    </tr>
                    <tr>
                        <td>密码</td>
```

```html
            <td>
                <input type="password" name="txtPassword" id="txtPassword"
                    width="500px" value="<%=info.getPassword()%>" />
            </td>
        </tr>
        <tr>
            <td>Email</td>
            <td>
                <input type="text" name="txtEmail" width="500px"
                    value="<%=info.getEmail()%>" />
            </td>
        </tr>
        <tr>
            <td>身份</td>
            <td>
                <select name="selPower">
                    <option value="admin">管理员</option>
                    <option value="user">一般用户</option>
                </select>
            </td>
        </tr>
        <tr>
            <td colspan="2">
                <input type="submit" name="button" id="button" value="提交" />
                <input type="reset" name="button2" id="button2" value="重置" />
                <input type="hidden" name="action" value="<%=pageAction%>" />
                <input type="hidden" name="username" value="<%=info.getUsername()%>" />
            </td>
        </tr>
    </table>
</form>
</body>
```

16. 后台用户删除页 users-delete.jsp

users-delete.jsp 位于 WebRoot\manager 目录，核心代码如下：

```jsp
<%@ page language="java" import="java.util.*" pageEncoding="utf-8"%>
<%@include file="islogin.jsp"%>
<%@page import="com.ch7.common.DataConverter"%>
<%@page import="com.ch7.dal.Users"%>
<%
    Users user = new Users(); //创建对象
    int result = 0;
    result = user.delete(request.getParameter("username"));
    if(result == 1) {
        out.println("<script>alert('用户删除成功');window.location.href('users-manage.jsp');</script>");
    } else {

        out.println("<script>alert('用户删除失败');window.location.href('users-manage.jsp');</script>");
```

```
        }
%>
```

7.4.4 前台表示层功能实现

1. 首页 index.jsp

其核心代码如下：

```jsp
<%@ page language="java" import="java.util.*" pageEncoding="utf-8"%>
<%@page import="com.ch7.model.BlogInfo"%>
<%@page import="com.ch7.dal.Blog"%>
<%@page import="com.ch7.common.DataValidator"%>
<%@page import="com.ch7.common.Utility"%>
<%@page import="com.ch7.dal.Class"%>
<%@page import="com.ch7.model.ClassInfo"%>

<%
Blog blog = new Blog();
List<BlogInfo> list = blog.getList(null);
//获得博文分类
Class cls = new Class();
List<ClassInfo> clist = cls.getList();
%>
...
<!-- start content -->
    <div id="content">
    <%
             for(BlogInfo info : list) { //遍历输出list集合中的数据
         %>
        <div class="post">
            <h1 class="title"><a href="blog-information.jsp?id=<%=info.getId()%> "><%=info.getTitle()%></a></h1>
            <p class="byline"><small><%=info.getCreatedtime()%> </small></p>
            <div class="entry">
                <p><%=Utility.Substring(DataValidator.removeHtml(info.getContext()),300) %></p>
            </div>
            <p class="meta"><a href="#" class="more">分类: <%=info.getClassName()%></a>    <a href="#" class="more">详情</a>    <a href="blog-information.jsp?id=<%=info.getId()%>#comment" class="comments">我要评论</a></p>
        </div>
     <%
             }
         %>
    </div>
    <!-- end content -->
...
<!-- start sidebar -->
    <div id="sidebar">
```

```
            <ul>
                <li>
                    <h2>日志分类</h2>
                    <ul>
                    <%for(ClassInfo cinfo : clist) {%>
                        <li><a href="blog-list.jsp?classId=<%=cinfo.getId()%>">
<%=cinfo.getName()%></a></li>
                    <%}%>
                    </ul>
                </li>
            </ul>
        </div>
        <!-- end sidebar -->
```

2. 博文详细页 blog-information.jsp

其核心代码如下：

```jsp
<%@ page language="java" import="java.util.*" pageEncoding="utf-8"%>
<%@page import="com.ch7.model.BlogInfo"%>
<%@page import="com.ch7.dal.Blog"%>
<%@page import="com.ch7.common.DataValidator"%>
<%@page import="com.ch7.common.Utility"%>
<%@page import="com.ch7.dal.Class"%>
<%@page import="com.ch7.model.ClassInfo"%>
<%@page import="com.ch7.common.DataConverter"%>
<%@page import="com.ch7.model.CommentInfo"%>
<%@page import="com.ch7.dal.Comment"%>
<%
int id = DataConverter.toInt(request.getParameter("id"));
Blog blog = new Blog();
BlogInfo info = blog.getBlogInfo(id);
if(info == null) {
    out.println("<script>alert('博文ID有误');window.location.href('index.jsp');</script>");
}
%>
...
<!-- start content -->
    <div id="content">
        <div class="post">
            <h1 class="title"><%=info.getTitle() %></h1>
            <p class="byline"><small><%=info.getCreatedtime()%> </small></p>
            <div class="entry">
                <p><%=DataValidator.serverHtmlDecode(info.getContext()) %></p>
            </div>
            <p class="meta"><a href="#" class="more">分类: <%=info.getClassName()%>
</a> </p>
        </div>
        <a name="comment" id="comment"></a>

    <p class="meta">评论: </p><br>
    <%
        Comment comment = new Comment();
        List<CommentInfo> cmlist =comment.getListByBlogId(id);
```

```
         for(CommentInfo cminfo : cmlist) {
            out.println(cminfo.getContext() +"<br>" );
            out.println(cminfo.getCreatedtime() +"    "+
cminfo.getUsername() +"<br><hr>" );
         }
     %>

     <jsp:include page="comment.jsp">
     <jsp:param value="<%=id%>" name="id"/>
     </jsp:include>
     </div>
     <!-- end content -->
```

3. 博文分类查找页 blog-list.jsp

其核心代码如下：

```
<%@ page language="java" import="java.util.*" pageEncoding="utf-8"%>
<%@page import="com.ch7.model.BlogInfo"%>
<%@page import="com.ch7.dal.Blog"%>
<%@page import="com.ch7.common.DataValidator"%>
<%@page import="com.ch7.common.Utility"%>
<%@page import="com.ch7.dal.Class"%>
<%@page import="com.ch7.model.ClassInfo"%>
<%@page import="com.ch7.common.DataConverter"%>
<%
Blog blog = new Blog();
int classId = DataConverter.toInt(request.getParameter("classId"));
List<BlogInfo> list = blog.getListByClassId(classId);
//获得博文分类
Class cls = new Class();
List<ClassInfo> clist = cls.getList();
 %>
 ...
<!-- start content -->
    <div id="content">
     <%
            for(BlogInfo info : list) { //遍历输出list集合中的数据
         %>
       <div class="post">
         <h1 class="title"><a href="blog-information.jsp?id=<%=
info.getId()%>"><%=info.getTitle()%></a></h1>
         <p class="byline"><small><%=info.getCreatedtime()%> </small></p>
         <div class="entry">
            <p><%=Utility.Substring(DataValidator.removeHtml
(info.getContext()),300) %></p>
         </div>
         <p class="meta"><a href="#" class="more">分类: <%=info.getClassName()%>
</a>   <a href="#" class="more">详情</a>     </p>
       </div>
       <%
           }
        %>
    </div>
    <!-- end content -->
```

4. 博文评论页 comment.jsp

其核心代码如下：

```jsp
<%@ page language="java" import="java.util.*" pageEncoding="utf-8"%>
<%@page
    import="com.ch7.common.Utility"%>
    <%@page import="com.ch7.common.DataValidator"%>
    <%@page import="com.sun.corba.se.spi.orbutil.fsm.Action"%>
    <%@page import="com.ch7.dal.Users"%>
    <%@page import="com.ch7.common.MD5"%>
    <%@page import="com.ch7.common.DataConverter"%>
    <%@page import="com.ch7.dal.Comment"%>
    <%@page import="com.ch7.model.CommentInfo"%>
<%
if("login".equals(request.getParameter("action")))
{
    Users users = new Users();
    String username = request.getParameter("txtUserName");
    String password = MD5.Encrypt(request.getParameter("txtPassword"));
    if(!users.isExist(username, password)) {
        out.println(" <script>alert('用户名密码有误');window.location.href='blog-information.jsp?id="+DataConverter.toInt(request.getParameter("id"))+"');</script>");
    } else {
        Utility.writeCookie(response, "user", username);
response.sendRedirect("blog-information.jsp?id="+DataConverter.toInt(request.getParameter("id")));
    }
}
if("save".equals(request.getParameter("action")))
{
    Comment comment = new Comment();
    CommentInfo info = new CommentInfo();
    info.setBlogid(DataConverter.toInt(request.getParameter("id")));
    info.setContext(request.getParameter("txtContext"));
    info.setUsername(Utility.readCookie(request,"user"));
    comment.insert(info);
    response.sendRedirect("blog-information.jsp?id="+DataConverter.toInt(request.getParameter("id")));
}
%>
<%
    String data = Utility.readCookie(request, "user");

    if(DataValidator.isNullOrEmpty(data)) {

%>
<form id="form1" name="form1" method="post" action="comment.jsp?action=login">
    <table width="400" border="0">
    <tr>
       <td>用户名</td><input name="id" type="hidden" id="id" value="<%=request.getParameter("id")%>"/>
```

第7章 综合实例——BLOG 系统开发

```jsp
        <td><input name="txtUserName" type="text" id="txtUserName" size="12" /></td>
        <td>密码</td>
        <td><input name="txtPassword" type="password" id="txtPassword" size="12" /></td>
        <td><input type="submit" name="button" id="button" value="提交" /> <a href="register.jsp">注册</a></td>
      </tr>
    </table>
  </form>
<%}else{ %>
  欢迎您：<%=Utility.readCookie(request,"user") %> <a href="logout.jsp">退出</a>
  <form name="commentForm" method="post" action="comment.jsp?action=save">
  <input name="id" type="hidden" id="id" value="<%=request.getParameter("id")%>"/>
        <textarea rows="4" cols="50" name="txtContext"></textarea>
        <input type="submit" name="button" value="回复">
      </form>
<%}%>
```

5. 会员注册 register.jsp

其核心代码如下：

```jsp
<%@ page language="java" import="java.util.*" pageEncoding="utf-8"%>
<%@page import="com.ch7.model.BlogInfo"%>
<%@page import="com.ch7.common.DataValidator"%>
<%@page import="com.ch7.common.Utility"%>
<%@page import="com.ch7.dal.Class"%>
<%@page import="com.ch7.model.ClassInfo"%>
<%@page import="com.ch7.model.UsersInfo"%>
<%@page import="com.ch7.dal.Users"%>
<%@page import="com.ch7.common.MD5"%>
<%  request.setCharacterEncoding("utf-8");
    String action = request.getParameter("action");
    UsersInfo info = new UsersInfo();
    Users user = new Users();
    if("reg".equals(action)) {
        if(user.isExistUsersInfo(request.getParameter("txtUserName")))
            out.println("<script>alert('用户已存在');window.location.href('register.jsp');</script>");
        else
        {
            info.setUsername(request.getParameter("txtUserName"));
            info.setPassword(MD5.Encrypt(request.getParameter("txtPassword")));
            info.setEmail(request.getParameter("txtEmail"));
            info.setPower("user");
            user.insert(info);
            out.println("<script>alert('注册成功');window.location.href('index.jsp');</script>");
        }
    }
%>
...
<!-- start content -->
```

```html
            <div id="content">
                <form id="form1" name="form1" method="post"
                    action="register.jsp?action=reg">
                    <table width="100%" border="0">
                        <tr>
                            <td>
                                用户名
                            </td>
                            <td>
                                <input type="text" name="txtUserName" id="txtUserName"
                                    width="500px" />
                            </td>
                        </tr>
                        <tr>
                            <td>
                                密码
                            </td>
                            <td>
                                <input type="password" name="txtPassword" id="txtPassword"
                                    width="500px" />
                            </td>
                        </tr>
                        <tr>
                            <td>
                                确认密码
                            </td>
                            <td>
                                <input type="password" name="txtPassword2" id="txtPassword2"
                                    width="500px" />
                            </td>
                        </tr>
                        <tr>
                            <td>
                                Email
                            </td>
                            <td>
                                <input type="text" name="txtEmail" width="500px" />
                            </td>
                        </tr>
                        <tr>
                            <td colspan="2">
                                <input type="submit" name="button" id="button" value="注册" />
                                <input type="reset" name="button2" id="button2" value="重置" />
                            </td>
                        </tr>
                    </table>
                </form>
            </div>
            <!-- end content -->
```

6. 会员退出 logout.jsp

其核心代码如下：

```jsp
<%@page language="java" pageEncoding="utf-8"%>
<%@page import="com.ch7.common.Utility"%>
<%@page import="com.ch7.common.DataValidator"%>
<%
    String data = Utility.readCookie(request, "user");
    if(DataValidator.isNullOrEmpty(data)) {
        response.sendRedirect("index.jsp");
    } else {
        Utility.writeCookie(response, "user", data, 0);
        response.sendRedirect("index.jsp");
    }
%>
```

7.5 系统运行界面

7.5.1 前台界面

前台界面如图 7-14～图 7-17 所示。

图 7-14 博客系统首页

图 7-15 博客系统博文详细页

图 7-16 博客系统会员注册页　　　　　图 7-17 博客"关于我"页面

7.5.2 后台界面

后台界面如图 7-18～图 7-27 所示。

图 7-18 后台登录窗口　　　　　　　　图 7-19 博文管理界面

图 7-20 博文编辑界面　　　　　　　　图 7-21 博文评论界面

第 7 章 综合实例——BLOG 系统开发

图 7-22 博文添加界面

图 7-23 博文分类管理界面

图 7-24 博文分类编辑界面

图 7-25 博文分类添加界面

图 7-26 用户管理界面

图 7-27 用户编辑界面

小 结

本章提供了一个基于 JSP+JavaBean 方式实现的博客系统。

在这个博客系统中，JavaBean 用于连接数据库以执行业务逻辑，或者作为值对象在与 JSP 传递数据，JSP 调用 JavaBean 执行业务逻辑。

通过本章学习，读者将熟悉 JSP+JavaBean 模式的 Web 开发方法。

习 题

新建一个工程 Ex7,参照本章博客系统的开发模式,设计并开发一个简易的在线考试系统。

提 高 篇

第 8 章
Struts2 入门

学习目标
- 了解 Struts2 框架。
- 掌握 Struts2 框架包的下载和环境配置。
- 熟悉 Struts2 的简单开发流程及应用。

Struts2 框架是基于 WebWork 来设计自己的架构，其解决了 Struts1 目前比较严重的缺陷，并推出了多个稳定版本，加之其能和 spring、hibernate 等优秀框架友好集成，目前已成为比较流行的 Web 开发框架之一，被广泛应用的趋势正在逐步上升。基于此，本章节开始对其进行一些基础的介绍和简单的应用。初学的读者可以把本章作为 Struts2 入门的开始，使用过此框架的读者则可以利用本章节巩固一些 Struts2 框架的基础知识。

8.1 Struts2 框架介绍

Struts2 是基于 Struts1 和 WebWork 的框架，虽然其是在 Struts1 的基础上发展起来的，但其却是以 WebWork 为核心设计思想的，同时，Struts2 也是诸多优秀 MVC 开源框架中的一员。

8.1.1 Struts1 概述

Struts1 自从第一个版本发布以来，不仅市场占有率高，而且用它来开发的人群也非常广泛。

Struts1 的控制器就是其核心（大部分 Web MVC 框架都是这样），其控制器由核心控制器和业务逻辑控制器两部分组成。Struts1 的核心控制器是 ActionServlet，由 Struts1 框架提供。Struts1 的业务逻辑控制器是由开发者提供的自定义的 Action。然而，随着 Web 技术的不断发展和 Web 应用范围的不断扩大，Struts1 也展现出其不符合目前发展的需求：

1．所支持的表现层比较单一

Struts1 所支持的表现层技术只有 JSP，并没有对目前很流行的 Velocity、FreeMarker、XSLT 等表现层技术提供支持。这一点使 Struts1 的应用受到了严重的制约。

2．与 Servlet API 严重耦合，测试困难

Struts1 的业务逻辑控制器内充满大量的 Servlet API。这一点可以体现出 Struts1 完全是基于 Servlet API，两者严重耦合。而 Servlet API 通常由 Web 服务器来实例化，所以 Struts1 对 Web 服务器的依赖性非常大，一旦脱离了 Web 服务器，其 Action 的测试将变得非常困难。

3．属于侵入式设计，不利于代码重用

Struts1 的 Action 中包含了大量的 Struts1 API，这种侵入式的设计不利于代码重用，一旦发生系统重构，这种侵入式的测试将变得非常困难。

8.1.2 MVC 概述

MVC 英文全称是 Model-View-Controller，中文翻译为模型-视图-控制器。这种设计模式在目前的 Web 开发中被广泛使用。MVC 设计模式的思想就是把 Web 应用程序分为三个核心部分：模型（Model）、视图（View）、控制器（Controller）。其结构如图 8-1 所示。

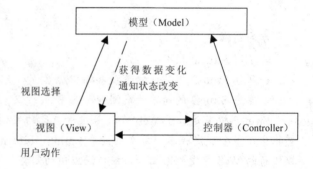

图 8-1　MVC 结构

（1）模型。模型是应用程序的主体部分。模型表示业务数据，或者业务逻辑。

（2）视图。视图是应用程序中用户界面相关的部分，是用户看到并与之交互的界面。

（3）控制器。控制器工作就是根据用户的输入，控制用户界面数据显示和更新 model 对象状态。例如：在网页单击"提交"按钮，把请求传输到控制器去处理。控制器接到请求后先调用模型处理业务数据，然后通知视图读取处理数据，并把结果显示出来，从而完成用户请求。

其处理过程是：用户在视图提供的界面上发出请求，视图将用户的请求转发给控制器。控制器调用相应的模型处理用户的请求。模型处理完业务逻辑后，把处理的结果返回给控制器。最终由控制器选择对应的视图把数据显示在界面上。

MVC 作为一种模块化的设计思想，给开发者带来了很多的方便性：

（1）多个视图能共享一个模型。在 MVC 设计模式中，模型响应用户请求并返回响应数据，视图负责格式化数据并把它们呈现给用户，业务逻辑和数据表示分离，同一个模型可以被不同的视图重用，所以大大提高了模型层程序代码的可重用性。

（2）模型是自包含的，与控制器和视图保持相对独立，因此可以方便地改变应用程序的业务数据和业务规则。如果把数据库从 MySQL 移植到 Oracle，或者把 RDBMS 数据源改变成 LDAP 数据源，只需改变模型即可。一旦正确地实现了模型，不管业务数据来

自数据库还是 LDAP 服务器，视图都会正确地显示它们。由于 MVC 的三个模块相互独立，改变其中一个不会影响其他两个，所以依据这种设计思想能构造良好的松耦合的组件。

（3）控制器提高了应用程序的灵活性和可配置性。控制器可以用来连接不同的模型和视图去完成用户的需求，控制器为构造应用程序提供了强有力的重组手段。给定一些可重用的模型和视图，控制器可以根据用户的需求选择适当的模型进行业务逻辑处理，然后选择适当的视图将处理结果显示给用户。

8.1.3 WebWork 概述

WebWork 于 2003 年 3 月正式发布，对 Struts1 框架进行了大幅度的改进，引入了当时不少新的概念、思想和功能，然而，其却和原来的 Struts1 框架不兼容，但其解决了 Struts1 所展现出来的缺陷和局限性。

8.1.4 Struts2 概述及优势

Struts2 框架兼容了 Struts1 和 WebWork 两个框架，继承了 WebWork 的优秀设计思想，摒弃了 Struts1 的缺陷和局限性，是市场和技术的结合，更加符合了 MVC 的设计思想，方便了开发人员，满足了市场，其使用人群正在逐步上升。下面是 Struts2 框架简单的处理流程：

（1）浏览器发送请求，例如 register.action。

（2）进入 web.xml 文件，其配置的核心控制器 FilterDispatcher 会根据请求决定调用合适的 Action。

（3）进入 Struts.xml 文件，Struts2 的拦截器链自动对请求应用通用功能，例如：WorkFlow、Validation 等功能。

（4）如果 Struts.xml 文件中配置 Method 参数，则调用 Method 参数对应的 Action 类中的 Method 方法，否则调用通用的 Execute 方法来处理用户请求。

（5）将 Action 类中的对应方法返回的结果响应给浏览器。

8.2 Struts2 的环境配置

要想使用 Struts2 框架进行 Web 开发或者运行以 Struts2 框架为基础的程序，首先要对 Struts2 框架包进行下载。虽然大部分开发软件会自带 Struts2 框架包，但也只是 Struts2 框架包的核心部分，而且没有与之相应的开发文档，所以为了更好地学习 Struts2 框架，有必要下载完整的 Struts2 框架包。

为了使读者对 Struts2 的开发环境配置细节有个感性的认识，本书在 Struts2 开发环境配置时对 Struts2 的开发环境完全进行手动配置，不采取软件导入 Struts2 的方式。

8.2.1 下载 Struts2 框架包

Struts 的官方网站是 http://struts.apache.org/，单击页面右上角的 struts2，进入 Struts2 框架包下载页面。在编写本章时，Struts2 框架包的最新版本为 Struts-2.3.4，其下载网址为 http://struts.apache.org/download.cgi#struts234，下载时有多个选项可供选择，本章选择 Full Distribution 选项，即 Struts2 的完整版。为了读者更好地学习 Struts2 框架，建议读者

与本书下载一致的框架包。

8.2.2 搭建 Struts2 开发环境

搭建 Struts2 的完整开发环境的步骤如下所示，依次进行即可。在本书的前几个章节有对 JDK、Tomcat、MySQL、Navicat_for_MySQL 和 MyEclipse 的安装及应用进行过详细介绍，这里不再赘述。

（1）安装 JDK。

（2）安装 Tomcat、MySQL 和 MyEclipse，这三者不分先后。

（3）安装数据库管理软件 Navicat_for_MySQL_10.0.11.0。

（4）将 JDK 和 Tomcat、MySQL 与 MyEclipse 相关联。

（5）先将上面下载的 Struts2 框架包进行解压，然后找到 lib 文件夹，之后将文件夹中的 struts2-core-2.3.4.jar、xwork-2.3.4.jar、ognl-3.0.5.jar、freemarker-2.3.19.jar、commons-logging-1.1.1.jar、commons-io-2.0.1.jar、commons-fileupload-1.2.2.jar、commons-lang3-3.1.jar 和 javassist-3.11.0.GA.jar 等常用 JAR 包复制到 Myeclipse 中需要搭建 Struts2 开发环境项目的相应目录下，其目录为 XX/WebRoot/WEB-INF/lib（XX 为项目名）。

（6）打开 XX/WebRoot/WEB-INF/下的 web.xml 文件，对其配置 Struts2 的核心控制器 FilterDispatcher 和为其核心控制器 FilterDispatcher 建立映射。具体情况如下面的代码所示。

```xml
<?xml version="1.0" encoding="UTF-8"?>
<web-app version="3.0"
    xmlns="http://java.sun.com/xml/ns/javaee"
    xmlns:xsi="http://www.w3.org/2001/XMLSchema-instance"
    xsi:schemaLocation="http://java.sun.com/xml/ns/javaee
    http://java.sun.com/xml/ns/javaee/web-app_3_0.xsd">
<display-name></display-name>
<welcome-file-list>
<welcome-file>index.jsp</welcome-file>
</welcome-file-list>
<!-- 配置struts2的核心控制器FilterDispatcher -->
<filter>
    <filter-name>struts2</filter-name>
    <filter-class>
    org.apache.struts2.dispatcher.ng.filter.StrutsPrepareAndExecuteFilter
    </filter-class>
</filter>
<!-- 为FilterDispatcher建立映射 -->
<filter-mapping>
    <filter-name>struts2</filter-name>
    <url-pattern>/*</url-pattern>
</filter-mapping>
</web-app>
```

（7）在 XX/src 目录下创建 XML（Basic Templates）文件，命名为 struts.xml，此文件的名字可自行定义，但文件扩展名不能更改。建议把它命名为 struts.xml，这是程序员写程序时的命名习惯。并加入如下所示配置代码：

```xml
<?xmlversion="1.0"encoding="UTF-8"?>
<!DOCTYPEstrutsPUBLIC"-//Apache Software Foundation//DTD Struts
```

```
Configuration 2.1//EN"
   "http://struts.apache.org/dtds/struts-2.1.dtd">
<struts>
</struts>
```

下面对复制到 MyEclipse 中需要搭建 Struts2 开发环境的项目相应目录下的常用 JAR 包功能进行介绍。

（1）struts2-core-2.3.4.jar：Struts2 的核心库。

（2）xwork-2.3.4.jar：WebWork 的核心库，需要它的支持。

（3）ognl-3.0.5.jar：OGNL 表达式语言，Struts2 支持该 EL 表达式。

（4）freemarker-2.3.19.jar：表现层框架，定义了 Struts2 的可视组件主题。

（5）commons-logging-1.1.1.jar：日志管理。

（6）commons-io-2.0.1.jar：支持附件上传。

（7）commons-fileupload-1.2.2.jar：支持文件上传。

（8）commons-lang3-3.1.jar：对 JDK 中 java.Lang 的补充。

（9）javassist-3.11.0.GA.jar：字节码的类库。

8.3　第一个 Struts2 示例

本示例用于处理用户注册请求，用户在显示的注册界面填写用户信息，然后提交。Struts2 接收请求、处理请求，并把处理的结果返回给用户。为了尽量简单，在程序中没有去将用户信息存入数据库，只是进行了一个简单的判断。图 8-2 所示为第一个 Struts2 示例的工作流程图。

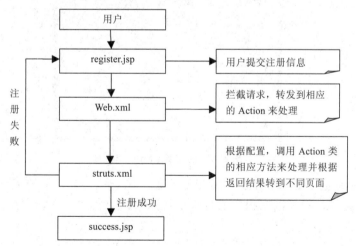

图 8-2　第一个 Struts2 示例工作流程

实现该 Struts2 应用程序，需要进行以下工作：

（1）视图模块：创建 register.jsp 用于显示用户注册的界面；创建 success.jsp 用于显示验证成功的界面。

（2）模型组件：本工程的逻辑简单，所以没有创建专门处理业务逻辑的 JavaBean，所有的逻辑处理工作在控制器 Action 类中完成。

（3）控制器组件：需要创建 Action 类处理表单数据，进行简单的逻辑判断，然后返回对应的值。

（4）配置文件：在 struts.xml 中添加代码，配置上面创建的 Action 类，指定访问它的 URL 以及根据其返回值调用的 JSP 页面；在 web.xml 中添加代码，使 FilterDispatcher 过滤所有的用户请求。

（5）表单验证：对用户输入的信息进行验证。本示例采用在 Action 类中添加方法去验证。

8.3.1 准备工作

用户注册示例的准备工作如下：

（1）完整 Struts2 开发环境配置的前四步，这里不再赘述，参考 8.2.2 小节。

（2）打开 MyEclipse，创建一个 Web Project，命名为 firstStruts2。具体如图 8-3 所示。

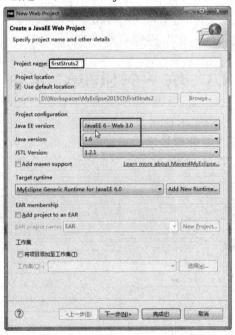

图 8-3　创建 firstStruts2 项目

（3）执行完整 Struts2 开发环境配置的第（5）~（7）步。把目录中的 XX 换成 firstStruts2，这样，就可以在相应的目录下找到相应的文件，完成配置。这里不再赘述，参考 8.2.2 小节。

8.3.2 配置 struts.xml 与 struts.properties 文件

Struts2 框架有两个核心配置文件：struts.xml 和 struts.properties。

（1）配置 struts.xml 文件，具体配置代码如下：

```
<?xml version="1.0" encoding="UTF-8" ?>
<!DOCTYPE struts PUBLIC "-//Apache Software Foundation//DTD Struts Configuration 2.1//EN"
```

```xml
    "http://struts.apache.org/dtds/struts-2.1.dtd">
<struts>
<!-- 设置Web应用程序的编码集 -->
<constant name="struts.i18n.endcoding" value="UTF-8" /><!-- 使用常量 -->
<!-- 设置Web应用程序的默认Locale -->
<constant name="struts.locale" value="zh_CN"></constant><!-- 使用常量 -->
<!-- 定义firstStruts2包,在这个包中定义Action类-->
<package name="firstStruts2" extends="struts-default">
    <!-- 定义register,用于处理用户注册请求 -->
    <action name="register" class="com.examp.ch18.RegAction1">
        <!-- 如果注册成功,返回success.jsp -->
        <result name="success">success.jsp</result>
        <!-- 如果输入数据有错误,返回register.jsp -->
        <result name="input">register.jsp</result>
    </action>
</package>
</struts>
```

可以看出，`<constant></constant>`标签设置了Web应用程序的编码集为UTF-8，本地语言为中文；在`<package></package>`之间定义控制器，一个控制器用一个`<action>`标签去配置，该标签的name属性指定访问该Action类的URL。Class用于指定Action的实现类，当Action类返回success字符串时，调用success.jsp去显示；当返回input时，调用register.jsp去显示。

（2）struts.properties是一个标准的Properties文件，该文件是由一系列key和value组成的，每个key就是一个Struts2的属性，该key对应的value就是一个Struts2的属性值。事实上，struts.properties文件的内容均可在struts.xml中以`<constant name="" value=""></constant>`加载（使用常量）。方便起见，没必要单独配置struts.properties文件，如果想建立struts.properties文件，通常放在XX/src目录下，在文件加载后，其会被加载到XX/WebRoot/WEB-INF/classes目录下。

8.3.3 创建控制器（Action类）

Struts2使用一个普通的Java类来接收用户请求，然后调用模型组件去处理业务逻辑，最后返回一个字符串，Struts2框架会根据这个字符串的值调用相应的页面去显示。在firstStruts2/src目录下建立com.examp.ch18，并在该包下建立处理用户注册请求的控制器RegAction1.java，其代码如下：

```java
package com.examp.ch18;
import java.util.Date;
import java.util.Map;
import com.opensymphony.xwork2.ActionContext;
public class RegAction1{
    private String username;// 定义用户名属性
    private String password1;// 定义密码属性
    private String password2;// 定义确认密码
    private Date birthday;// 定义生日属性
```

```java
    public String getUsername() {
        return username;
    }
    public void setUsername(String username) {
        this.username = username;
    }
    public String getPassword1() {
        return password1;
    }
    public void setPassword1(String password1) {
        this.password1 = password1;
    }
    public String getPassword2() {
        return password2;
    }
    public void setPassword2(String password2) {
        this.password2 = password2;
    }
    public Date getBirthday() {
        return birthday;
    }
    public void setBirthday(Date birthday) {
        this.birthday = birthday;
    }
    public String execute() throws Exception {
        Map m;// 定义 Map 类型变量 Map m;
        if(getPassword1().equals(getPassword2())
                && !getUsername().trim().equals("")) {
            m = ActionContext.getContext().getSession(); // 取得 session
            m.put("username", getUsername());// 将 name 存入 session
            return "success";
        } else {
            return "input";
        }
    }
}
```

可以看出，RegAction1 定义了属性 username、password1、password2、birthday，并定义了这些属性的 getter 和 setter 方法。setXXX()括号中的参数的名字必须和表单参数的名字一样。这样 Struts2 框架就可以把表单的值附给这些属性，并且表单也可以获取该类的属性值。在 execute()方法中处理用户请求，由于仅用于测试，没有调用模型层把用户信息存入数据库。

8.3.4 创建视图层

视图层是一些显示页面，因为目前还没学 Struts2 标签库，所以页面是以 html 为基础的。显示页面存在于 firstStruts2/WebRoot 目录下。

（1）注册页面 register.jsp。表单请求交给了 register.action，程序通过 web.xml 文件找到 struts.xml 中 name 属性名称为 register 的 action 标签，进而找到此动作的执行类 RegAction1，其中 register.jsp 页面代码如下：

```
<%@ page contentType="text/html; charset=UTF-8" language="java" %>
<!DOCTYPE html PUBLIC "-//W3C//DTD XHTML 1.0 Transitional//EN"
"http://www.w3.org/TR/xhtml1/DTD/xhtml1-transitional.dtd">
<html>
<body>
用户信息注册: <br><hr>
<form method="get" action="register.action">
<table>
<tr><td>姓名: <input name="username" type="text"></td></tr>
<tr><td>密码: <input name="password1" type="password"></td></tr>
<tr><td>验证密码: <input name="password2" type="password"></td></tr>
<tr><td>生日:<input name="birthday" type="text"></td></tr>
<tr><td><input type=submit value="注册"></td></tr>
</table>
</form>
</body>
</html>
```

（2）注册成功后的返回页面 success.jsp。success.jsp 通过${sessionScope.username}获取 RegAction1 类中设置在 session 中的值，代码如下：

```
<%@ page language="java" contentType="text/html; charset=UTF-8"%>
<!DOCTYPE html PUBLIC "-//W3C//DTD HTML 4.01 Transitional//EN"
"http://www.w3.org/TR/html4/loose.dtd">
<html>
<head>
<meta http-equiv="Content-Type" content="text/html; charset=utf-8">
<title>欢迎界面</title>
</head>
<body>
您好! ${sessionScope.username}，注册成功!
</body>
</html>
```

8.3.5 测试运行

把示例部署到 Tomcat 服务器上，然后打开 register.jsp 页面，从上到下依次在页面的文本框内输入 admin、123456、123456、1987-12-09，如图 8-4 所示。

单击"注册"按钮，因填写的信息符合验证条件，所以注册成功，页面转到 success.jsp，如图 8-5 所示。

如果在打开的 register.jsp 页面，从上到下依次在页面的文本框内输入 admin、12345、4567、1987-12-09，单击"注册"按钮，因为填写的两次密码不一样，即不符合验证条件，所以会注册失败，页面转到 register.jsp。

 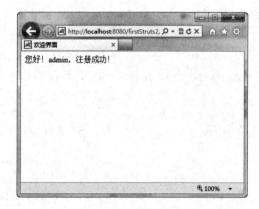

图 8-4 填写注册信息　　　　　　　图 8-5 注册成功

小　　结

　　本章首先通过对 Struts2 框架进行理论上的介绍,让读者对 Struts2 框架有一定的了解,认识到 Struts2 框架是目前比较符合现在发展需求的开发框架;接下来介绍了 Struts2 框架包的下载和完整开发环境的配置;然后,通过一个非常简单的用户注册示例向读者介绍了使用 Struts2 框架进行 Web 开发的基本流程、简单应用、重要文件配置,使读者对 Struts2 框架有了一个感性的认识。

习　　题

　　根据本章所讲内容,自行到网上下载 Struts2 的 JAR 包,将其配置到新建的 Web 项目中,并对根据书中所讲的标签进行验证总结。

第 9 章
Struts2 框架技术

学习目标

- 了解 Struts2 的标签库及部分标签的应用。
- 掌握 Struts2 的国际化操作及应用。
- 掌握 Struts2 的 validate()验证方法。
- 掌握 Struts2 的配置文件验证方法。

Struts2 框架提供了比较全面的标签库供开发者使用，掌握 Struts2 标签库是读者学习 Struts2 框架不可缺少的一部分。学会 Struts2 框架的国际化，有利于软件等的用户不再受地域限制，方便共享。同时，Struts2 框架的校验方式也比较多，熟练的掌握一两种 Struts2 框架的校验方法，有助于读者以后的学习和应用。本章是 Struts2 框架的提高篇。

9.1 Struts2 的标签库

Struts2 框架对其使用者提供了定义好的标签库，此标签库不仅功能强大，而且简单易用，大大简化了视图层的代码，提高了视图页面的开发维护效率。

Struts2 标签其实就是 Java 类生成基于 XML 的脚本的方法。一个 Struts2 标签代表着一系列复杂的 Java 代码，这些代码实现了该标签的功能。所以一个 Struts2 标签也可以看作一个可以重用的功能模块的入口，标签的参数也是功能模块的参数。Struts2 标签的使用极大地减少了 JSP 文件中 Java 脚本的使用量，使代码变得简洁。大量的标签组合可以形成一个标签库，一旦建立一个标签库，就可以在多个项目中使用它进行视图开发，标签库具有很强的通用性和重用性。

事实上，Struts2 标签库并没有把标签进行严格细致的分类，而是把所有的标签都融合到一个标签库里。本章为了使读者清晰地认识 Struts2 标签库，结合多方文献，将 Struts2 标签库大致分为三大类，并对这些大类进行了细分，具体如下所述：

（1）UI 标签：主要用来生成 HTML 页面元素。其又可以详细划分为两大类：表单标签，用来生产 HTML 页面中的 FORM 元素；非表单标签，用来生成页面上的 tree、tab 等。

（2）非 UI 标签：主要用来进行数据访问、逻辑控制等。其又可以详细划分为两大类：数据标签，主要用于对数据的处理和存储等；逻辑控制标签，用来实现分支、循环等流程控制。

（3）Ajax 标签：主要用来对 Ajax 技术的支持。

下面用图来表示 Struts2 标签库的分类，使读者有个整体印象，具体情况如图 9-1 所示。

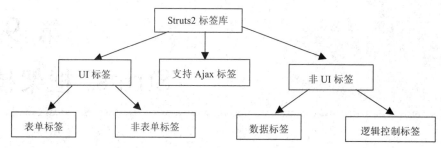

图 9-1 Struts2 标签库分类

Struts2 标签库是个非常庞大的体系，事实上，在开发应用中并不是每个标签都会用到，基于此，本章节只对经常用到的标签进行介绍。

9.1.1 Struts2 标签库的使用

使用 Struts2 标签库必须先指定标签定义文件（TLD 文件）位置，该文件记录了实现标签功能文件的位置、属性等。JSP 容器通过该文件获知在什么地方调用标签库。解压 struts2-core-2.3.4.jar 文件，在 META-INF 下就可以找到 Struts2 标签库的标签定义文件 struts-tag.tld。要在 JSP 文件中使用 Struts2 标签库中的标签，只需使用 taglib 编译指令导入标签库 <%@ taglib prefix="s" uri="/struts-tags" %>即可。下面用一个简单的表单标签演示下在 JSP 文件中使用 Struts2 标签的格式，具体代码如下：

```
<%@ page contentType="text/html;charset=UTF-8" language="java"%>
<%@ taglib prefix="s" uri="/struts-tags"%>
<html>
    <head>
        <title>Struts2 标签格式简单示例！</title>
    </head>
    <body>
    <s:form action=""></s:form>
    </body>
</html>
```

9.1.2 if/elseif/else 标签

if/elseif/else 这三个标签是属于 Struts2 的非 UI 标签的逻辑控制标签。if 标签主要用于控制选择输出，elseif 和 else 标签都是和 if 标签结合使用的，这三者结合在一起，跟多数编程语言中的 if/elseif/else 语句的功能相似，主要用于进行程序的分支逻辑控制。if/elseif/else 标签的使用格式如下：

```
<%@ page contentType="text/html;charset=UTF-8" language="java"%>
<%@ taglib prefix="s" uri="/struts-tags"%>
<html>
```

```
    <head>
        <title>if/elseif/else 标签示例</title>
    </head>
    <body>
        <s:set name="testname" value="%{'Java'}" />
        <s:if test="%{#testname=='Java'}">
            <div><s:property value="%{#testname}" /></div>
        </s:if>
        <s:elseif test="%{#testname=='Jav'}">
            <div><s:property value="%{#testname}" /></div>
        </s:elseif>
        <s:else>
            <div>testname 不是"Java"</div>
        </s:else>
    </body>
</html>
```

上述代码的行顺序为：set 标签定义了一个名为 testname 的属性，并为属性赋了初始值，初始值为 Java；if/elseif/else 标签根据代码中所给定的 testname 的属性值进行控制输出，并通过<s:property value="%{#testname}" />语句对符合条件的属性值进行输出。此代码运行结果在页面输出为 Java。同时要注意到 if 和 elseif 标签必须指定 test 的属性，并且要和 set 标签的 name 属性值保持一致，因为 if 和 elseif 标签的属性是用来获得逻辑表达式返回的真假值的。

9.1.3 iterator 标签

iterator 标签属于 Struts2 的非 UI 标签的逻辑控制标签，主要用来对集合属性进行遍历输出，其中的集合属性可能是 List、Map、Set 或者是数组。下面通过实例讲解。

创建一个 Action 类，定义一个 List 属性，一个 Map 属性。编写该类：iteratorTag.java，具体代码如下：

```java
package com.examp.ch18;

import java.util.ArrayList;
import java.util.HashMap;
import java.util.List;
import java.util.Map;

import com.opensymphony.xwork2.ActionSupport;

@SuppressWarnings({ "serial", "rawtypes" })
public class iteratorTag extends ActionSupport {
    private List myList;
    private Map myMap;
    @SuppressWarnings("unchecked")
    public String execute() throws Exception {
        myList = new ArrayList();
        myList.add("第一个元素");
        myList.add("第二个元素");
```

```java
        myList.add("第三个元素");

        myMap = new HashMap();
        myMap.put("key1", "第一个元素");
        myMap.put("key2", "第二个元素");
        myMap.put("key3", "第三个元素");
        return SUCCESS;
    }
    public List getMyList() {
        return myList;
    }
    public void setMyList(List myList) {
        this.myList = myList;
    }
    public Map getMyMap() {
        return myMap;
    }
    public void setMyMap(Map myMap) {
        this.myMap = myMap;
    }
}
```

从上面代码可以看出,创建了 myList、myMap 这 2 个属性,每个属性都存储了 3 个元素,用于在 JSP 页面中显示。打开 struts.xml 文件,在名为 firstStruts2 的包内添加 action 元素,代码如下:

```xml
<action name="iteratorTag" class="com.examp.ch18.iteratorTag">
    <result name="success">iteratorTag.jsp</result>
</action>
```

下面创建 iteratorTag.jsp 页面,该文件用 iterator 标签遍历输出,具体代码如下所示:

```jsp
<%@ page contentType="text/html;charset=UTF-8" language="java"%>
<%@ taglib prefix="s" uri="/struts-tags"%>
<html>
    <head>
        <title>Iterator 标签示例! </title>
    </head>
    <body>
        <h1>
            <span style="background-color: #aaaabb">Iterator 标签示例</span>
        </h1>
        <h2>显示 List</h2>
        <table>
            <s:iterator value="{'第一个元素','第二个元素'}" status="st">
                <tr>
                    <td><s:property value="#st.getIndex()" /></td>
                    <td><s:property /></td>
                </tr>
            </s:iterator>
```

```
            </table>
            <h2>显示 List 属性</h2>
            <table>
                <s:iterator value="myList" status="st">
                    <tr>
                        <td><s:property value="#st.getIndex()" /></td>
                        <td><s:property /></td>
                    </tr>
                </s:iterator>
            </table>
            <h2>显示 Map</h2>
            <table>
                <s:iterator value="#{'key1':'第一个元素','key2':'第二个元素'}" status="st">
                    <tr>
                        <td><s:property value="#st.getIndex()" /></td>
                        <td><s:property /></td>
                    </tr>
                </s:iterator>
            </table>
            <h2>显示 Map 属性</h2>
            <table>
                <s:iterator value="myMap" status="st">
                    <tr>
                        <td><s:property value="#st.getIndex()" /></td>
                        <td><s:property /></td>
                    </tr>
                </s:iterator>
            </table>
        </body>
</html>
```

在 MyEclipse 中执行该文件，最后的执行结果如图 9-2 所示。

通过实例的编写和运行，读者对 iterator 标签的 value 属性有了一定的了解，事实上，其还有 status、id 等两个可选属性，这三个属性的具体情况如下：

（1）id 属性：可选属性，指定了集合的 id。

（2）value 属性：可选属性，该属性指定要进行遍历的集合对象。例如{'第一个元素','第二个元素'}就是一个集合，myList 也是一个集合，它是 Action 类的属性。

（3）status 属性：可选属性，该属性是每次遍历时都会产生一个的 IteratorStatus 实例对象，例如：

图 9-2　Iterator 标签示例执行结果

#st.getIndex()就是调用 IteratorStatus 实例的 getIndex()方法。以下是 iteratorStatus 实例对

象的常用方法，见表 9-1。

表 9-1 iteratorStatus 实例对象的常用方法

方法	描述
int getIndex()	返回当前被遍历元素的索引值
int getCount()	返回已经遍历元素的总数
Boolean isEven()	判断当前迭代元素是否为偶数
Boolean is Odd()	是否为奇数
Boolean is First()	是否为集合中第一个元素
Boolean isLast()	是否为集合中最后一个元素

9.1.4 include 标签

include 标签属于非 UI 标签的数据标签。include 标签主要用于在 JSP 页面里包含其他资源，例如，JSP 文件、Servlet 等资源文件，当项目用到的页面较多，并且各个页面中有许多一样的代码，把重复率高的代码单独做成文件，运用此标签将这类文件引入相应的文件位置里，会节省好多代码，也会提高开发效率。此标签有 id 和 value 两个属性，其主要作用如下：

（1）id 属性：是可选属性，主要用于指定该标签的引用 id，一般情况下用得比较少。

（2）value 属性：是必选属性，主要用来引入 JSP 和 Servlet 等资源文件。

下面通过一个实例来介绍 include 标签的使用，需要创建被引入的文件 inc.jsp，主文件 include.jsp。引入文件 inc.jsp，具体代码如下：

```
<%@page language="java" import="java.util.*" pageEncoding="utf-8"%>
<!DOCTYPEHTMLPUBLIC"-//W3C//DTD HTML 4.01 Transitional//EN">
<html>
<head>
    <title>inc.jsp 页面</title>
</head>
<body>
    <h1>这是被引入文件 inc.jsp 页面</h1>
    您的登录名为：<%out.print(request.getParameter("name")); %>
</body>
</html>
```

上述代码是通过 request 对象的 getParameter()方法得到传递的参数，并通过页面输出语句 out.print 输出内容。因为 inc.jsp 文件得到的是主文件 include.jsp 文件的请求参数，要想获得参数就必须用 request.getParameter("name")手工调用。主文件 include.jsp 的代码如下：

```
<%@page language="java" import="java.util.*" pageEncoding="utf-8"%>
<%@taglib prefix="s" uri="/struts-tags"%>
<!DOCTYPEHTMLPUBLIC"-//W3C//DTD HTML 4.01 Transitional//EN">
<html>
<head>
    <title>include.jsp</title>
</head>
<body>
```

```
        <s:include value="inc.jsp">
        <s:param name="name" value="'管理员'"></s:param>
        </s:include>
    </body>
</html>
```

上述代码中，通过使用<s:include></s:include>标签将 inc.jsp 文件引入主文件 include.jsp 中，并通过 param 标签将 include.jsp 里面 name 的参数和参数值传递给被引入文件 inc.jsp。程序的运行结果如图 9-3 所示。

图 9-3　include.jsp 运行结果

9.1.5　property 标签

property 标签属于非 UI 标签的数据标签。property 标签的具体功能是输出 value 属性指定的值，此标签有 id、value、escape 和 default 等四个属性，它们的具体功能如下：

（1）id 属性：用于指定该元素的引用 id，用的情况比较少。

（2）value 属性：用于指定需要输出的属性值，经常要用。

（3）escape 属性：用于指定标签代码是否显示，默认值为 true，不显示的话则指定属性值为 false。

（4）default 属性：当属性值为 null 时，用来指定输出的值。

property 标签的应用实例在 9.1.3 小节中已经使用过，其使用方法这里不再赘述。

9.1.6　部分 UI 标签的使用

UI 标签主要是针对于 HTML 页面的使用，主体上大致分为表单标签和一些常用功能标签，下面通过一个注册实例来介绍 UI 标签的使用，具体情况见如下代码。因为此案例要用到一些日期等功能，所以要到 8.2.1 小节下载的框架包里的 lib 目录下找到开发所需的 struts2-dojo-plugin-2.3.4.jar 文件，并将其复制到配置好的 Struts2 的项目 XX/WebRoot/WEB-INF/lib 目录下（例如，第 8 章，本书配置过 firstStrus2 项目，在本节可以将 struts2-dojo-plugin-2.3.4.jar 文件复制到 firstStrus2/WebRoot/WEB-INF/lib 目录下）。新建一个 register1.jsp 文件，其代码如下：

```
<%@ page contentType="text/html; charset=UTF-8" pageEncoding="UTF-8" %>
<%@ taglib prefix="s" uri="/struts-tags" %>
<%@ taglib prefix="sx" uri="/struts-dojo-tags" %>
<html>
<head>
<title>注册实例</title>
<sx:head/>
</head>
<body>
<s:form action="register.action"><%-- 表单 --%>
    <table>
        <s:textfield label="账号" name="userName"/><%-- 文本框 --%>
```

```
                <s:password label="密码" name="password"/><%-- 密码 --%>
                <s:password label="确认密码" name="repassword"/>
                <sx:datetimepicker tooltip="在这里输入你的生日" label="出生日期"
name="birthday" /><%-- 日期--%>
                <s:textarea tooltip="输入你的个人介绍" label="个人简介" name="bio"
cols="20" rows="3"/><%-- 文本域--%>
                <s:select tooltip="性别" label="性别" list="{'男', '女'}"
name="sex"/><%-- 下拉列表--%>
                <s:checkboxlist tooltip="选择你最喜欢的明星" label="偶像"
                    list="{'刘德华', '陈奕迅', '黄家驹', '成龙', '李连杰'}"
name="friends"/><%-- 复选框--%>
                <s:radio list="{'文学类', '技术类', '其他'}" name="enjoy" label="喜
欢的书籍"/><%-- 单选框--%>
                <s:optiontransferselect
                        tooltip="选择你最喜欢的卡通人物 "
                        label="最喜欢的卡通人物"
                        name="leftSideCartoonCharacters"
                        leftTitle="左栏"
                        rightTitle="右栏"
                        list="{'奥特曼', '柯南', '超人'}"
                        multiple="true"
                        headerKey="headerKey"
                        headerValue="--- 请选择 ---"
                        emptyOption="true"
                        doubleList="{'孙悟空', '米老鼠', '唐老鸭'}"
                        doubleName="rightSideCartoonCharacters"
                        doubleHeaderKey="doubleHeaderKey"
                        doubleHeaderValue="--- 请选择 ---"
                        doubleEmptyOption="true"
                        doubleMultiple="true" /><%-- 移动标签--%>
                <s:submit value="注册"/><%-- 注册--%>
        </table>
    </s:form>
    </body>
</html>
```

从上面的代码可以看出，运用标签能够减少代码的书写量，使整个的页面代码看起来简洁、清晰。但是在实际的应用项目开发中会发现，标签不是很灵活，无法完成某些特殊的功能，在此采取的一种折中方案是：把标签穿插在原来的 HTML 页面设计中，既保证代码的灵活性，又可使用封装好的标签减少编写代码的工作量。

UI 标签使用起来并不复杂，大多数标签的属性都很相似，并且大多数的标签能和 HTML 的代码编写对应，理解起来不复杂。本实例囊括了大部分 UI 标签，对读者的具体应用有一定的帮助。本小节不准备对每个 UI 标签都进行细致讲解，只对标签的应用实践和技巧进行重点讲解，以方便大家入门。例如，大部分 UI 标签都有 label 属性，其为标签在页面显示的主题内容，如，账号行，label 的代码为 label="账号"，其在页面的显示内容为"账号"；name 属性为提交内容的字段名；有 list 词的属性一般为列举内容；很多的

标签属性可以根据 HTML 的代码编写情况对照出来，用法没有改变。运行页面如图 9-4 所示。

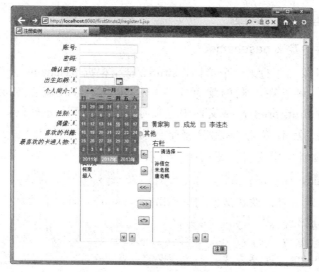

图 9-4 注册实例运行结果界面

9.2 Struts2 的国际化操作

国际化是这样定义的：在程序不做任何修改的情况下和在不同的国家、地区以及不同的语言环境下，按照本地机器的语言和地区设置来显示相应的字符。例如，一个中文网站被其他国家的用户访问时，就需要按照本地机器的语言和地区设置来显示相应的页面。Struts2 框架提供了这样的种国际化操作接口，用户只需要编写代表不同语言的资源文件并进行相关配置，就可以实现国际化的效果。

9.2.1 Struts2 实现国际化的原理

Struts2 的国际化是基于 Java 的国际化，其只是优化和封装了 Java 的国际化，使国际化的实现过程更简单了些。图 9-5 是 Struts2 国际化的运行流程图，主要是为了便于读者熟悉 Struts2 的国际化原理。

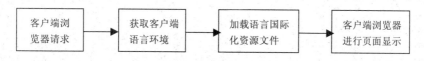

图 9-5 Struts2 国际化运行流程图

当客户端浏览器发出请求后，Struts2 的 il8n 拦截器对客户端浏览器的请求进行拦截，并获取客户端浏览器的地区语言环境的 request_locale 的值，获得 request_locale 的值后，il8n 拦截器将其实例化程 Locale 对象，并存储在用户相应的 Session 对象中。接下来，Struts2 在 struts.xml 或者 struts.properties 配置文件中查找相应国际化文件并进行加载（struts.xml 和 struts.properties 的关系在 8.3.2 小节中讲解过，这里不再赘述）。加载好国际化资源文件后，Struts2 的视图文件会通过 Struts2 标签把国际化的信息在客户端浏览器显示出来。

9.2.2 实现国际化步骤

1. 在配置文件中定义 basename

basename 是资源文件的第一个单词，struts2 框架根据 basename 去查找资源文件。其为一个常量定义，可以在不同的配置文件中去实现。假设现在要定义 basename 为 globalMessage，在 struts.properties 文件中定义时，要加如下所示代码：

```
struts.custom.i18n.resources=globalMessages
```

在 struts.xml 文件中定义时，要加如下所示代码（本小节实例中采取这种方式）：

```
<constant name="struts.custom.i18n.resources" value="globalMessages">
```

在 web.xml 中定义时，要在文件中的<filter>元素下添加如下所示代码：

```
<init-param>
    <param-name> struts.custom.i18n.resources</param-name>
    <param-value>globalMessages</param-value>
</init-param>
```

配置好 basename 后，开始编写资源文件，其内容由键（key）—值（value）对组成，其命名格式为 basename_language_country.properties，例如，中文资源文件可以用如下格式来命名：globalMessages_zh_CN.properties，英文资源文件可以命名为：globalMessages_en_US.properties。

2. 编写中英文资源文件

定义 globalMessages_zh_CN.properties，资源文件代码如下所示：

```
user=用户名称
pass=用户密码
username=注册用户名
password1=密码
password2=确认密码
birthday=生日
regpage=注册界面
errpage=错误界面
successlogin=登录成功
falselogin=登录失败
regsuccess=注册成功
regfalse=对不起,注册失败
regbutton=注册
```

然后，再定义一个名为 globalMessages_en_US.properties 的属性文件，这是 struts2 国际化的英文环境的资源属性文件，代码如下：

```
user=username
pass=password
username=Your Name
password1=Password
password2=confirm Password
```

```
birthday=Birthday
regpage=Reg Page
errpage=ERROR Page
successlogin=Welcom
falselogin=Sorry!You can't log in
regsuccess=OK,You reg success!
regfalse=Sorry! You Reg False!
regbutton=Reg!
```

3．将中文资源文件转换为 unicode 编码

由于中文资源文件中出现了非西欧字符，必须把中文字符转换为 unicode 码才能被 Struts2 识别，这是因为 Java 的国际化是通过 unicode 编码实现的。JDK 提供了一个 native2ascii 转换工具，在 DOS 环境下，将中文资源文件中的中文字符转化为 unicode 编码。执行此过程的步骤是：

（1）单击电脑"开始"菜单，在搜索框中 cmd，然后按【Enter】键，弹出窗口如图 9-6 所示。

（2）切换用户的操作目录，进入到存放 globalMessages_zh_CN.properties 文件的目录下，例如 globalMessages_zh_CN.properties 存放在 D:/目录下，则在命令行输入"D:"然后按【Enter】键就可切换到 D 盘的根目录下，如图 9-7 所示。

图 9-6 cmd 命令窗口

图 9-7 切换用户的操作目录

（3）输入 native2ascii 命令，将 globalMessages_zh_CN.properties 文件转化为 unicode 编码的文件 new_globalMessages_zh_CN.properties，命令的输入方法如图 9-8 所示。

图 9-8 运行 native2ascii 命令

（4）执行完该命令后，在用户的当前目录下就会多出了一个名为 new_globalMessages_zh_CN.properties 的属性文件，该文件就是将 globalMessages_zh_CN.properties 文件中的内容用 unicode 进行编码生成的，用 MyEclipse 将该文件打开，看到该文件的内容如下：

```
user=\u7528\u6237\u540d\u79f0
pass=\u7528\u6237\u5bc6\u7801
username=\u6ce8\u518c\u7528\u6237\u540d
password1=\u5bc6\u7801
password2=\u786e\u8ba4\u5bc6\u7801
birthday=\u751f\u65e5
regpage=\u6ce8\u518c\u754c\u9762
errpage=\u9519\u8bef\u754c\u9762
successlogin=\u767b\u5f55\u6210\u529f
falselogin=\u767b\u5f55\u5931\u8d25
regsuccess=\u6ce8\u518c\u6210\u529f
regfalse=\u5bf9\u4e0d\u8d77,\u6ce8\u518c\u5931\u8d25
regbutton=\u6ce8\u518c
```

（5）转换成功后，将 new_globalMessages_zh_CN.properties 中的代码放入 global Messages_zh_CN.properties 中。然后将 globalMessages_zh_CN.properties 与 globalMessages_en_US.properties 一起保存在 firstStruts2/src 目录下（firstStruts2 为第 8 章建立的项目）。

4．用户注册页面

创建使用 key 值的用户注册页面 register2.jsp，将其存放在 firstStruts2/WebRoot 目录下。页面具体代码如下所示：

```
<%@ page contentType="text/html;charset=UTF-8" language="java" %>
<%@ taglib prefix="s" uri="/struts-tags" %>
<%@ taglib prefix="sx" uri="/struts-dojo-tags" %>
<html>
<head>
<title><s:text name="regpage"/></title><%--引入资源文件中的 regpage --%>
<sx:head />
</head>
<body>
<table>
<s:form id="id" action="Reg.action">
<s:textfield name="username" key="username"/>
<s:password name="password1" key="password1"/>
<s:password
    name="password2"
    key="password2"/>
<sx:datetimepicker
    name="birthday"
    key="birthday" />
<s:submit key="regbutton"/>
</s:form>
</table>
</body>
</html>
```

5. 编写 Action 类

Action 类也可以使用特定的方法根据 key 值查找对应的 value 值。在 firstStruts2 项目的 com.examp.ch18 包下创建 Reg.java 类。在 execute()方法中获取 value 值并打印输出，代码如下：

```java
package com.examp.ch18;
import java.util.Date;
import com.opensymphony.xwork2.ActionSupport;
public class Reg extends ActionSupport {
    private static final long serialVersionUID = -3436886820441245650L;
    // 定义用户名属性
    private String username;
    // 定义处理信息：注意同 http 中的 msg 不同名称
    private String mymsg;
    // 定义密码属性
    private String password1;
    // 定义确认密码
    private String password2;
    // 定义生日属性
    private Date birthday;
    public String execute() throws Exception {
        if(username != null && getPassword1().equals(getPassword2())
                && !getUsername().trim().equals("")) {
                //使用 getText 方法获取参数对应的 Value 值
            System.out.println(getText("username") + ":" + username);
            System.out.println(getText("password1") + ":" + password1);
            System.out.println(getText("birthday") + ":" + birthday);
            return SUCCESS;
        } else {
            return INPUT;
        }
    }
    public String getUsername() {
        return username;
    }
    public void setUsername(String username) {
        this.username = username;
    }
    public String getMymsg() {
        return mymsg;
    }
    public void setMymsg(String mymsg) {
        this.mymsg = mymsg;
    }
    public String getPassword1() {
        return password1;
    }
    public void setPassword1(String password1) {
        this.password1 = password1;
    }
    public String getPassword2() {
```

```java
        return password2;
    }
    public void setPassword2(String password2) {
        this.password2 = password2;
    }
    public Date getBirthday() {
        return birthday;
    }
    public void setBirthday(Date birthday) {
        this.birthday = birthday;
    }
}
```

可以看出，通过 getText(Key)方法获得资源文件中的 value 值，并用该值作为用户输入值的标签打印在控制台上。

6. 编写注册成功的 success1.jsp 页面

将其存放在 firstStruts2/WebRoot 目录下。此页面用来显示用户注册成功的界面，同样也要国际化，代码如下：

```jsp
<%@ page contentType="text/html;charset=UTF-8" language="java" %>
<%@ taglib prefix="s" uri="/struts-tags" %>
<html>
<head>
  <title><s:text name="regsuccess"/></title>
  <s:head />
</head>
<body>
  <table>
    <h2><s:text name="username"/>: <s:property value="username" /></h2>
    <h2><s:text name="password1"/>: <s:property value="password1" /></h2>
    <h2><s:text name="birthday"/>: <s:property value="birthday" /></h2>
  </table>
</body>
</html>
```

上面的文件没有使用 key=""去访问资源文件，而是使用了 struts2 标签中的<s:text>标签来实现对资源文件的访问。

7. 对 struts.xml 进行配置

增加一个 Action，指明其具体的实现类为 com.examp.ch18.Reg，并有两个逻辑视图分别为 "success" 和 "input"，详细的配置代码如下所示。将除了 <constant name="struts.custom.i18n.resources" value="globalMessages"/>代码以外的<constant>标签代码去掉（如留下来会影响测试）。

```xml
<action name="Reg" class="com.examp.ch18.Reg">
        <result name="success">success1.jsp</result>
        <result name="input">register2.jsp</result>
</action>
```

8. 运行测试

通过在浏览器中切换语言环境进行测试。浏览器切换语言环境的步骤为：

第 9 章　Struts2 框架技术

（1）单击浏览器导航栏的工具栏目中 Internet 选项，如图 9-9 所示。

（2）在弹出的对话中单击最下方的"语言"按钮，打开"语言首选项"对话框。单击添加按钮进行语言选择，选择"英国（美国）[en-US]"选项，单击"确定"按钮，如图 9-10 所示。选中"语言"区域内任意语言选项进行上移或者下移，改变语言的首选项，之后单击"确定"按钮即可。

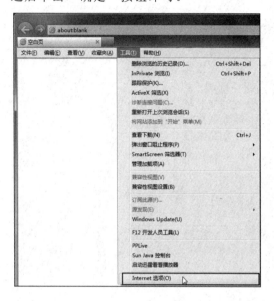
图 9-9　Internet 选项路径

图 9-10　"语言首选项"对话框

图 9-11 为中文环境下的国际化输出，图 9-12 为英文（美国）环境下的输出，其注册成功后进入 succes1.jsp 界面输出与语言环境也是一致的，这里就不再使用图片说明。

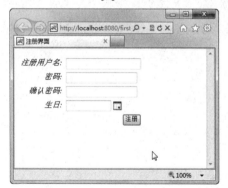
图 9-11　中文环境下的国际化输出

图 9-12　英文（美国）环境下的国际化输出

9.3　Struts2 数据验证

9.3.1　使用 validate()方法进行验证

1．编写 Action 类

在编写 Action 类时，一般会继承 ActionSupport 类，此类提供了一些规范的静态常量，

并且提供了对表单数据进行验证的方法 validate()。我们对 8.3 小节中实例的 RegAction1.java 类进行重新的改写，就可以对表单进行验证了。下面新建一个 RegAction2.java 类，将其存放在 firstStruts2 项目的 com.examp.ch18 包下，代码如下：

```java
package com.examp.ch18;

import java.util.Date;
import com.opensymphony.xwork2.ActionSupport;

@SuppressWarnings("serial")
public class RegAction2 extends ActionSupport {
    private String username;        //定义用户名属性
    private String password1;       //定义密码属性
    private String password2;       //定义确认密码
    private Date birthday;          //定义生日属性
    public String getUsername() {
        return username;
    }
    public void setUsername(String username) {
        this.username = username;
    }
    public String getPassword1() {
        return password1;
    }
    public void setPassword1(String password1) {
        this.password1 = password1;
    }
    public String getPassword2() {
        return password2;
    }
    public void setPassword2(String password2) {
        this.password2 = password2;
    }
    public Date getBirthday() {
        return birthday;
    }
    public void setBirthday(Date birthday) {
        this.birthday = birthday;
    }
    public String execute() throws Exception {
        return SUCCESS;
    }
    public void validate() {//数据校验方法
        //检查用户名是否为空
        if(username == null || username.trim().equals("")) {
            //写入校验错误
            addFieldError("username", "用户名不能为空！");
        }
        //检查生日不能为空
        if(birthday == null || birthday.after(new Date())) {
```

```
            //写入校验错误
            addFieldError("birthday", "生日日期不正确");
        }
        //检查密码是否为空
        if(password1 == null || getPassword1().trim().equals("")) {
            //写入校验错误
            addFieldError("password1", "密码不能为空！");
        }
        //确认密码不一致
        if(!getPassword1().equals(getPassword2())) {
            //写入校验错误
            addFieldError("password2", "两次输入密码不一样！");
        }
    }
}
```

可以看出这个类比 RegAction1.java 的区别在于：

（1）在 execute()方法中，return SUCCESS 是一个 ActionSupport 中的常量，使用它可以规范代码，防止不同的程序员在表示成功时返回不同的字符串。

（2）多了一个 validate()方法，对用户输入信息进行验证，如果有错误，将错误信息显示在注册页面上。

（3）有了 validate()方法，execute()方法中的逻辑就变得很简单，不用再去进行判断。

2．创建 register3.jsp 页面

将其存放 firstStruts2/WebRoot 目录下，register3.jsp 页面代码如下：

```
<%@ page language="java" import="java.util.*" pageEncoding="utf-8"%>
<%@ taglib prefix="s" uri="/struts-tags" %>
<!DOCTYPE HTML PUBLIC "-//W3C//DTD HTML 4.01 Transitional//EN">
<html>
    <head>
        <title>注册页面</title>
    </head>
    <body>
        <table align="center" width="40%">
            <tr>
            <td style="color:red">
            <s:fielderror></s:fielderror>
            </td>
            </tr>
        </table>
        <form action="Reg2.action" method="post">
            <table align="center" width="40%" border="1">
                <tr>
                    <td>
                        用户名
                    </td>
                    <td>
                        <input type="text" name="username" value="${requestScope.username}">
```

```html
                            </td>
                        </tr>
                        <tr>
                            <td>
                                密码
                            </td>
                            <td>
                                <input type="password1" name="password1">
                            </td>
                        </tr>
                        <tr>
                            <td>
                                确认密码
                            </td>
                            <td>
                                <input type="password2" name="repassword2">
                            </td>
                        </tr>
                        <tr>
                            <td>
                                生日
                            </td>
                            <td>
                                <input type="text" name="birthday">
                            </td>
                        </tr>
                        <tr>
                            <td>
                                <input type="submit" value=" 注册 ">
                            </td>
                            <td>
                                <input type="reset" value=" 重置 ">
                            </td>
                        </tr>
                    </table>
            </form>
        </body>
</html>
```

success.jsp 文件使用 8.3 小节实例中的 success.jsp 文件。struts.xml 文件中加入 <constant></constant>标签的代码<constant name="struts.locale" value="zh_CN"></constant>，在 struts.xml 文件的<package></package>标签内加入如下配置代码：

```
<action name="Reg2" class="com.examp.ch18.RegAction2">
    <result name="success">success.jsp</result>
    <result name="input">register3.jsp</result>
</action>
```

部署项目，运行后部分页面的运行结果如图 9-13 所示。

第 9 章 Struts2 框架技术

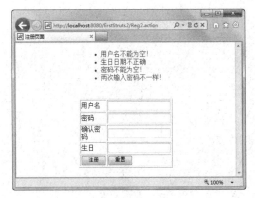

图 9-13 提交数据都为空时的验证

9.3.2 使用配置文件进行验证

除了使用 validate() 方法进行验证，还可以使用编写 XML 文件进行某个 action 对应表单的验证。这个文件的命名规则为 ActionName-validation.xml，并且要保存在和 Action 类相同的目录下。例如，Reg.java 类的验证文件为 Reg-validation.xml。示例对用户注册进行验证，将其保存为 RegXML.java，代码如下：

```java
package com.examp.ch18;
import java.util.Date;
import com.opensymphony.xwork2.ActionSupport;

@SuppressWarnings("serial")
public class RegXML extends ActionSupport {
    private String username;        //定义用户名属性
    private String password1;       //定义密码属性
    private String password2;       //定义确认密码
    private Date birthday;          //定义生日属性
    private int age;

    public String execute() throws Exception {
        return SUCCESS;
    }
    public String getUsername() {
        return username;
    }
    public void setUsername(String username) {
        this.username = username;
    }
    public String getPassword1() {
        return password1;
    }
    public void setPassword1(String password1) {
        this.password1 = password1;
    }
```

```java
    public String getPassword2() {
        return password2;
    }
    public void setPassword2(String password2) {
        this.password2 = password2;
    }
    public Date getBirthday() {
        return birthday;
    }
    public void setBirthday(Date birthday) {
        this.birthday = birthday;
    }
    public int getAge() {
        return age;
    }
    public void setAge(int age) {
        this.age = age;
    }
}
```

再新建一个 RegXML-validation.xml，此文件的参考格式在本书下载是 Struts2 框架包中能找到，绝对路径为 struts2\struts-2.3.4-all\struts-2.3.4\apps\struts2-mailreader\WEB-INF\src\java\mailreader2。其文件代码如下：

```xml
<?xml version="1.0" encoding="UTF-8"?>
<!DOCTYPE validators PUBLIC "-//Apache Struts//XWork Validator 1.0.2//EN"
"http://struts.apache.org/dtds/xwork-validator-1.0.2.dtd">
<validators>
    <!--
    第一个校验 field: username
    -->
    <field name="username">
        <field-validator type="stringlength">
            <param name="minLength">4</param>
            <param name="maxLength">10</param>
            <message>框架校验：用户名称长度为${minLength}到${maxLength}之间</message>
        </field-validator>
        <field-validator type="requiredstring" short-circuit="false">
            <message>框架校验：用户名不能为空！</message>
        </field-validator>
    </field>
    <field name="password1">
        <field-validator type="requiredstring">
            <message>框架校验：密码不能为空！</message>
        </field-validator>
        <field-validator type="stringlength">
            <param name="minLength">4</param>
```

```xml
            <param name="maxLength">20</param>
            <message>框架校验：用户密码长度为${minLength}到${maxLength}之间</message>
        </field-validator>
    </field>
    <field name="password2">
        <field-validator type="requiredstring">
            <message>框架校验：密码不能为空！</message>
        </field-validator>
        <field-validator type="stringlength" >
            <param name="minLength">4</param>
            <param name="maxLength">20</param>
            <message>框架校验：用户密码长度为${minLength}到${maxLength}之间</message>
        </field-validator>
        <field-validator type="fieldexpression">
            <param name="expression">password1==password2</param>
            <message>框架校验：两次输入密码不同！</message>
        </field-validator>
    </field>
    <!--
    校验field: birthday
    -->
    <field name="birthday">
        <!--
        定义类型为date
        -->
        <field-validator type="date">
            <!--
            校验生日有效时间段
            -->
            <param name="min">1990-01-01</param>
            <param name="max">2006-01-01</param>
            <message>框架校验：生日数据错误！</message>
        </field-validator><!--comment-->
    </field>
    <field name="age">
        <field-validator type="int">
            <param name="min">10</param>
            <param name="max">100</param>
            <message>框架校验：年龄必须为${min}为${max}之间！</message>
        </field-validator>
    </field>
</validators>
```

　　页面设计和 struts.xml 的配置是大致一样的，只需根据实际情况改动，这里不再赘述。学习程序语言首先要学会分析别人的代码，本小节把页面设计和 struts.xml 的配置留给读者，给读者一个思维锻炼和学习的机会。

小 结

  本章介绍了一些比较常用的 Struts2 框架的标签，比较详细地介绍了 Struts2 框架的国际化，同时也介绍了 Struts2 框架的两种校验方式，使读者进一步认识了 Struts2 框架，并对其基础应用有了一定的了解。

习 题

  新建一个 Web 工程 Ex9，参考本章内容，开发一个简单的会员登录系统，要把所学的 Struts2 框架的标签应用到的系统中，对会员的登录进行验证，并对系统进行国际化设置。

第 10 章
Hibernate 概述及实例分析

学习目标
- 了解 Hibernate 的作用及意义。
- 掌握 Hibernate 的安装配置方法。
- 掌握 Hibernate 配置文件和映射文件的书写规则。
- 掌握 Hibernate 的基本映射。
- 掌握使用 Eclipse 开发 Hibernate 应用的方法。

Hibernate 是一个开放源代码的对象关系映射框架,它对 JDBC 进行了非常轻量级的对象封装,使得程序员不需要和复杂的 SQL 语句打交道,只要像平时操作对象一样可以随心所欲地使用对象编程思维来操纵数据库。Hibernate 可以应用在任何使用 JDBC 的场合,既可以在 Java 的客户端程序使用,也可以在 Servlet/JSP 的 Web 应用中使用。它为面向对象的领域模型到传统的关系型数据库的映射,提供了一个使用方便的框架。Hibernate 也是目前 Java 开发中最为流行的数据库持久层框架。

Hibernate 的设计目标是将软件开发人员从大量相同的数据持久层相关编程工作中解放出来。无论是从设计草案还是从一个遗留数据库开始,开发人员都可以采用该框架。Hibernate 不仅负责从 Java 类到数据库表的映射(还包括从 Java 数据类型到 SQL 数据类型的映射),还提供了面向对象的数据查询检索机制,从而极大地缩短了手动处理 SQL 和 JDBC 的开发时间。

10.1 Hibernate 框架介绍

10.1.1 持久化和 ORM 简介

数据持久化就是将内存中的数据模型转换为存储模型,以及将存储模型转换为内存中的数据模型的统称。数据模型可以是任何数据结构或对象模型(通常情况下都是 JavaBean 对象),存储模型可以是关系模型、XML、二进制流等。由于对象模型和关系模型应用广泛,所以不少人会误认为数据持久化就是对象模型到关系型数据库的转换,实际上 Hibernate 只是对象模型到关系模型之间转换的一个具体实现。

ORM(Object-Relational Mapping,对象关系映射),是随着面向对象的信息系统开发方法发展而产生的。面向对象的开发方法是目前企业级应用开发环境中的主流开发方法,关系数据库则是应用最为广泛的主流数据存储系统。在软件开发过程中对象和关系数据具有两种不同的表现形式，一种在内存中表现为对象，另一种在数据库中表现为关系数据。内存中对象与对象之间存在关联和继承关系，而在数据库中，关系数据无法直接表达关联和继承关系。因此，为了解决对象与关系数据不匹配的现象，对象关系映射技术应运而生。对象—关系映射（ORM）系统一般以中间件的形式存在，主要实现程序对象到关系数据库数据的映射。其主要作用是把对象模型，例如 JavaBean 对象和关系型数据库的表建立对应关系，并且提供了一个通过 JavaBean 对象去操作数据库表的机制，从而实现以面向对象的方式来操作关系型数据库，提高开发的效率。

10.1.2 Hibernate 框架

Hibernate 是一个开放源代码的对象关系映射框架，它对 JDBC 进行了非常轻量级的对象封装，使得 Java 程序员可以随心所欲地使用对象编程思维来操纵数据库。Hibernate 的目标是释放开发者通常的数据持久化相关的编程任务的 95%。Hibernate 可以应用在任何使用 JDBC 的场合，既可以在 Java 的客户端程序使用，也可以在 Servlet/JSP 的 Web 应用中使用，最具革命意义的是，Hibernate 可以在应用 EJB 的 Java EE 架构中取代 CMP，完成数据持久化的重任。

Hibernate 是一种强大的可提供对象—关系持久化和查询服务的中间件，它可以使程序员依据面向对象的原理开发持久化类，实现对象之间的关联、继承、多态、组合、集合等。

Hibernate 提供了它特有的数据库查询语言 HQL，这种查询语言屏蔽了不同数据库之间的差别，使用户可以编写统一的查询语句执行查询。不同于其他持久化解决方案的是 Hibernate 并没有把 SQL 的强大功能屏蔽掉，而仍然兼容 SQL，这使得以往的关系技术依然有效。

总之，Hibernate 不仅仅管理 Java 类到数据库表的映射（包括 Java 数据类型到 SQL 数据类型的映射），还提供数据查询和获取数据的方法，可以大幅减少软件开发时人工使用 SQL 和 JDBC 处理数据的时间。

10.2 Hibernate 的环境配置

10.2.1 下载 Hibernate 框架包

当要在 Java 项目中使用 Hibernate 时，需要登录其官方网站，下载 Hibernate 的安装包，然后配置到项目中。目前，Hibernate 加入了 JBoss，而 JBoss 又加入了 Red Hat，因此登录 www.hibernate.org、www.jboss.org 或者 www.redhat.com 都可以看到 Hibernate 的链接，都可以下载 Hibernate 的发布版本。Hibernate 的最新版本是 4.1.5 版，其中 4.1.5 版本有 sp1 版和 Final 版之分，考虑到框架的稳定性问题，我们使用的版本是 Hibernate4.1.5.Final 版，如果读者想尝试使用最新的 Hibernate 功能可以从其官方网站上下载 sp1 版。但为了读者更好地学习 Hibernate 框架，建议读者使用与本书一致的框架包。从网站上下载 Hibernate 框架的压缩包 hibernate-release-4.1.5.Final.zip 压缩包，该压缩包不仅包含 Hibernate 的开发包，而且包含 Hibernate 编译和运行所依赖的第三方类库。

解压缩下载后的压缩包，解压缩后得到一个名为 hibernate-release-4.1.5.Final 的文件夹，该文件夹下有 documentation、lib 和 project 文件夹。Hibernate 框架所有用到的 JAR 文件都在 lib 文件夹下。在 lib 目录下还有 envers、jpa、optional 和 required 这 4 个文件夹，如果要使用 Hibernate 的基本功能只需要导入 lib 目录下 required 文件夹下的所有 JAR 包即可。Hibernate 框架还提供了一些实用的扩展功能，所需 JAR 文件都放置在另外 3 个文件夹中。但是如果一个 Java EE 项目只是使用到 Hibernate 的基本功能，通常不需要把所有的 JAR 包都加载进来。

10.2.2 搭建 Hibernate 开发环境

当使用 MyEclipse 开发工具来进行 Hibernate 应用程序的开发，Hibernate 的配置十分简单，因为 MyEclipse 开发工具对 Hibernate 做了很好地整合支持，通过简单的设置就可以对项目增加 Hibernate 框架支持。本节仍然以 MySQL 作为默认数据库。

（1）启动 MyEclipse，在要添加 Hibernate 支持的 Web 项目上右击，在弹出的快捷菜单中选择"构建路径"→"配置构建路径"命令，如图 10-1 所示。

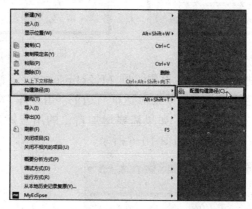

图 10-1　为 Web 项目配置构建路径

（2）选择"配置构建路径"命令后，出现如图 10-2 所示的项目属性对话框，该对话框主要用于设置编译和运行该项目所需的第三方类库。

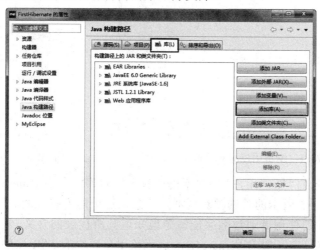

图 10-2　编辑 Web 项目构建路径的对话框

(3)选中"库"选项卡,单击右边的"添加库"按钮,在出现的对话框中选中"用户库",并单击"下一步"按钮,如图10-3所示,将出现如图10-4所示的选择用户库对话框,这个过程表示将选择一个用户库,一个用户库可以包含多个JAR文件。

图10-3 增加用户库对话框

图10-4 选择用户库对话框

如果读者是第一次进行用户库的设置,则MyEclipse中没有任何用户库,如图10-4所示,这就需要添加自己的用户库。单击对话框右边的"用户库"按钮,将出现如图10-5所示的编辑用户库的首选项对话框。如果需要增加自己的用户库,单击"新建"按钮,将出现输入用户库名字的对话框,如图10-6所示。

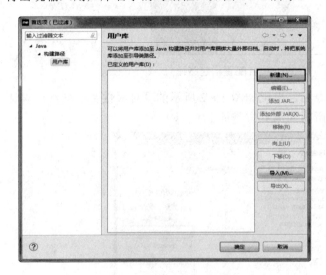

图10-5 编辑用户库对话框

图10-6 "新建用户"库对话框

在图10-6所示的"新建用户"库对话框中输入用户库的名字Hibernate4.1.5,单击"确定"按钮返回图10-5所示的首选项对话框。注意,这里的用户库的名字虽然可以任意命名,但为了用户库的使用者通过名字能清楚地知道其所代表的含义,用户库的命名要使用有一定意义的字符串,如Hibernate4.1.5表示该用户库中的JAR文件是Hibernate框架4.1.5版本的类库文件集合。另外,新建的用户库仅有一个用户库名,还没有包含任何的

第 10 章　Hibernate 概述及实例分析

JAR 文件，必须向用户库中添加所需的 JAR 文件。

选中图 10-5 所示对话框中需要编辑的用户库的名字，然后单击"添加外部 JAR"按钮，将出现一个文件浏览对话框，在该文件对话框中选择下载的 Hibernate 框架解压缩后的 lib/required 路径下的所有 JAR 文件，然后单击"打开"按钮，为 Hibernate4.1.5 用户库添加完成 JAR 文件后，看到用户库结构图如图 10-7 所示。

图 10-7　添加完 JAR 文件后 Hibernate4.1.5 用户库的结构图

单击图 10-7 所示中的"确定"按钮，返回如图 10-5 所示编辑用户库对话框。该界面中就新增了一个名字为 Hibernate4.1.5 用户库项。按照同样的方法，再分别增加几个常用的用户库：jakarta-commoning 用户库（jakarta 提供的一些通用功能的 JAR 包）、MySQL-Driver 用户库（mysql 数据库连接驱动的 jar 包）和 c3p0-0.9.1 用户库（c3p0 数据库连接池的 JAR 包），它们所包含的 JAR 文件如图 10-8 所示。

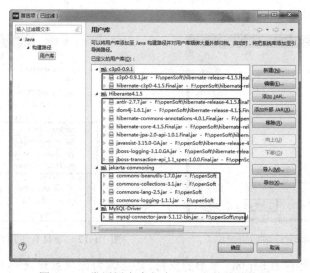

图 10-8　常用用户库包含 JAR 文件的结构图

把这些经常用到的 JAR 文件分门别类地放到不同的用户库下，便于开发编程人员加载这些整理好的 JAR 包，因为用户库里面的 JAR 文件在其他的项目开发中也会常常用到，这样在以后进行项目开发就不必再去寻找这些 JAR 文件，直接通过添加用户库到项目中即可。

单击"确定"按钮，返回到如图 10-9 所示编辑用户库对话框。

此步骤中，不要勾选刚才新建的"Hibernate4.1.5 用户库""jakarta-commoning 用户库""MySQL-Driver 用户库"和"c3p0-0.9.1 用户库"，单击"取消"按钮，返回如图 10-2 所示的对话框，单击对话框下面的"确定"按钮，返回 MyEclipse 的主界面。

右击该 Web 项目，在弹出的快捷菜单中选择"构建路径"→"添加库"命令，如图 10-10 所示。

图 10-9　不要勾选项目中的用户库

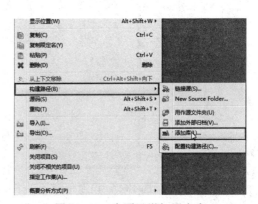

图 10-10　为项目增加用户库

随后，出现"添加库"对话框（见图 10-3），选择其中的"用户库"类型，并单击"下一步"按钮，将出现如图 10-11 所示的选择用户库对话框。

选中刚刚建立的这几个用户库，然后单击"完成"按钮，项目就增加了 Hibernate 框架类库的支持，看到的包资源管理器视图如图 10-12 所示，表明该 FirstHibernate 项目包含了新建立的 4 个用户资源库。

图 10-11　为项目增加用户库

图 10-12　项目有了用户资源库的支持

要使建立的项目有 Hibernate 框架的支持，还必须为该项目建立一个 Hibernate 配置文件，该 Hibernate 配置文件默认名称为 hibernate.cfg.xml，默认存放位置为该项目下的 src

第 10 章　Hibernate 概述及实例分析

目录下。具体操作步骤为：在项目上单击右击，在弹出的快捷菜单中选择"新建"→"文件"命令，操作过程如图 10-13 所示。

之后弹出如图 10-14 所示的"新建文件"对话框，在该对话框中输入存放 Hibernate 配置文件的父文件夹（配置文件 hibernate.cfg.xml 的存放路径），Hibernate 配置文件的文件名为 hibernate.cfg.xml。

图 10-13　建立 Hibernate 默认配置文件　　　图 10-14　建立 Hibernate 默认配置文件
　　　　　hibernate.cfg.xml　　　　　　　　　　　　　hibernate.cfg.xml 对话框

然后单击"完成"按钮，Hibernate 配置文件建立成功，将出现如图 10-15 所示的界面。

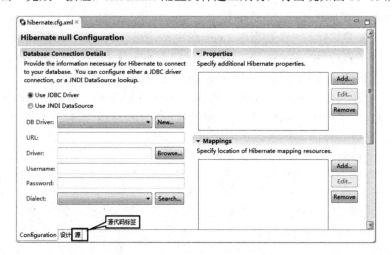

图 10-15　Hibernate 配置文件的可视化配置界面

如图 10-15 所示的界面是 Hibernate 配置文件的可视化配置界面，但该可视化配置界面的配置操作并不灵活，我们通常是通过源代码界面的方式对 hibernate.cfg.xml 文件进行配置的。单击如图 10-15 左下角的源代码标签，然后将下列源代码输入该 XML 配置文件中：

```xml
<?xml version='1.0' encoding='UTF-8'?>
<!DOCTYPE hibernate-configuration PUBLIC
        "-//Hibernate/Hibernate Configuration DTD 3.0//EN"
        "http://hibernate.sourceforge.net/hibernate-configuration-3.0.dtd">

<hibernate-configuration>
    <session-factory>
        <property name="hbm2ddl.auto">update</property>
        <property name="dialect">org.hibernate.dialect.MySQLInnoDBDialect
            </property>
        <property name="connection.url">jdbc:mysql://localhost:3306/javaee
            </property>
        <property name="connection.username">root</property>
        <property name="connection.password">1234</property>
        <property name="connection.driver_class">com.mysql.jdbc.Driver</property>
    </session-factory>
</hibernate-configuration>
```

该文件为 Hibernate 默认的配置文件,当程序调用 Configuration 对象的 configure()方法时,Hibernate 将自动加载该文件。Hibernate 配置文件的第一行是 XML 文件声明,指定该文件的版本和编码所用的字符集。Hibernate-configuration 是其根元素,根元素中有 session-factory 子元素,该元素依次有很多 property 元素,这些 property 元素配置 Hibernate 连接数据库的必要信息。其中,hbm2ddl.auto 的值为 update,表示 Hibernate 可根据 Hibernate 的映射文件自动创建数据表;dialect 表示使用的数据库方言,其值为 org.hibernate.dialect.MySQLInnoDBDialect;connection.url 表示数据库连接的 URL,其值为 jdbc:mysql://localhost:3306/javaee,即连接的数据库名称为 javaee,如果 MySQL 数据库中不存在名为 javaee 的数据库则需要新建这样的一个数据库。connection.username 和 connection.password 分别表示登录数据库服务器的用户名和密码,其值为分别是 root 和 1234;connection.driver_class 表示连接数据库使用的驱动类,其值为 com.mysql.jdbc.Driver。

值得注意的是,Hibernate 并不推荐采用 DriverManager 来连接数据库,而是推荐使用数据源来管理数据库连接,这样能保证最好的性能。Hibernate 推荐使用 C3P0 数据源,所以我们在上面配置文件的基础上添加配置数据源连接的信息,包括最大连接数、最小连接数等信息。完整的 Hibernate 配置文件如下(新增配置信息以斜体字显示):

```xml
<?xml version='1.0' encoding='UTF-8'?>
<!DOCTYPE hibernate-configuration PUBLIC
        "-//Hibernate/Hibernate Configuration DTD 3.0//EN"
        "http://hibernate.sourceforge.net/hibernate-configuration-3.0.dtd">

<hibernate-configuration>

    <session-factory>
        <property name="hbm2ddl.auto">update</property>
        <property name="dialect">org.hibernate.dialect.MySQLInnoDBDialect
            </property>
        <property name="connection.url">jdbc:mysql://localhost:3306/javaee
```

```
            </property>
        <property name="connection.username">root</property>
        <property name="connection.password">1234</property>
        <property name="connection.driver_class">com.mysql.jdbc.Driver</property>

        <property name="hibernate.c3p0.max_size">20</property>
        <property name="hibernate.c3p0.min_size">1</property>
        <property name="hibernate.c3p0.timeout">5000</property>
        <property name="hibernate.c3p0.max_statements">100</property>
        <property name="hibernate.c3p0.idle_test_period">3000</property>
        <property name="hibernate.c3p0.acquire_increment">2</property>
        <property name="hibernate.c3p0.validate">true</property>

    </session-factory>

</hibernate-configuration>
```

现在该 Web 项目就增加了 Hibernate 框架支持，软件编程人员可以利用 Hibernate 提供的功能进行数据持久化操作了。

10.3 第一个 Hibernate 示例

10.3.1 准备工作

下面开发第一个完整的 Hibernate 应用程序，通过该实例掌握 Hibernate 的基本使用方法。首先，启动 MyEclipse 工具，选择"新建"→"Web Project"命令，如图 10-16 所示。

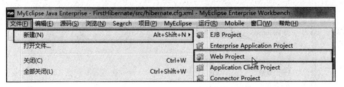

图 10-16 新建一个 Web 项目

弹出新建 Web 项目对话框，将该项目命名为 FirstHibernate，选择 Java EE 版本为 JaveEE6-Web3.0，然后单击"完成"按钮，新建项目成功，如图 10-17 所示。

为新建的 FirstHibernate 项目增加最新版本的 Hibernate 框架支持，其步骤读者请参看 10.2.2 节所介绍的内容来完成。等这些步骤完成后，在 MyEclipse 左边的导航窗口中出现如图 10-18 所示的导航树。

另外，为了验证 Hibernate 的功能，还需要在计算机上安装 MySQL 数据库，以及操作数据库的图形化客户端软件 Navicat_for_MySQL。至于这些软件的安装使用，前面章节已有详细介绍，这里不再赘述。

打开 Navicat_for_MySQL，新建一个名为 javaee 的数据库。注意这里的数据库的名称一定要与 Hibernate 配置文件的数据库的名称一致。如果不一致，将它们的名称修改成一致，否则数据库连接出错。新建的数据库 javaee 表中没有数据，如图 10-19 所示。

图 10-17　为新建的 Web 项目命名设置　　图 10-18　为 Web 项目添加 Hibernate 框架支持
　　　　　　　　　　　　　　　　　　　　　　　　　　JAR 包结构图

图 10-19　新建名为 javaee 的数据库

到此，准备工作全部完成。

10.3.2　创建 POJO 和 Hibernate 映射文件

下面在 FirstHibernate 项目中新建一个 News POJO 持久化类，具体步骤为：右击如图 10-18 所示的 FirstHibernate 结点，在弹出的快捷菜单中选择"新建"→"类"命令，出现如图 10-20 所示的新建类的对话框。

图 10-20　新建 POJO 类的对话框

输入该 POJO 类的名称为 News，该类所在的包为 com.chpt10.model，单击"完成"按钮，MyEclipse 就新建了一个名为 News.java 文件到该项目的 src 目录下。双击该 News.java 文件，为该 POJO 输入相应的属相和 setter 及 getter 方法，最终的文件源代码如下：

```java
package com.chpt10.model;

public class News {
    // 消息类的标识属性
    private int id;
    // 消息标题
    private String title;
    // 消息内容
    private String content;
    // 标识属性的 setter 和 getter 方法
    public void setId(int id) {
        this.id = id;
    }
    public int getId() {
        return (this.id);
    }
    // 消息标题的 setter 方法和 getter 方法
    public void setTitle(String title) {
        this.title = title;
    }
    public String getTitle() {
        return (this.title);
    }
    // 消息内容的 setter 方法和 getter 方法
    public void setContent(String content) {
```

```
            this.content = content;
    }
    public String getContent() {
        return (this.content);
    }
}
```

下面要创建 POJO 持久化类对应的 Hibernate 映射文件。再次右击 FirstHibernate 结点下的 com.chpt10.model 子结点,在弹出的快捷菜单中选择"新建"→"XML(Basic Templates)"命令,出现如图 10-21 所示的"新建 XML 文件"对话框。

图 10-21 "新建 XML 文件"对话框

在如图 10-21 所示的对话框中,填写文件名为 News.hbm.xml,其中 News 对应刚刚新建的 POJO 持久化类 News,后面的.hbm.xml 是固定写法,表明该文件是 Hibernate 的 XML 格式的映射文件。单击"完成"按钮,可以看到新生成的 News.hbm.xml 文件,该文件除了第一行的 XML 声明信息还没有内容,添加 POJO 持久化类的映射信息,最后的 XML 文件源代码为:

```xml
<?xml version="1.0" encoding="UTF-8"?>
<!DOCTYPE hibernate-mapping PUBLIC
    "-//Hibernate/Hibernate Mapping DTD 3.0//EN"
    "http://hibernate.sourceforge.net/hibernate-mapping-3.0.dtd">
<!-- hibernate-mapping 是映射文件的根元素 -->
<hibernate-mapping>
        <!-- 每个 class 元素对应一个持久化对象 -->
        <class name="com.chpt10.model.News" table="news_table">
            <!-- id 元素定义持久化类的标识属性 -->
            <id name="id" type="int" column="news_id">
                <generator class="identity"/>
            </id>
            <!-- property 元素定义普通属性 -->
            <property name="title" type="string"/>
            <property name="content"/>
```

第 10 章　Hibernate 概述及实例分析

```
        </class>
</hibernate-mapping>
```

　　该文件是 Hibernate 的映射文件，映射文件的 2～4 行指定了 Hibernate 映射文件的 DTD 信息，这几行对所有的 Hibernate 的映射文件都是相同的。<hibernate-mapping.../>元素是所有 Hibernate 映射文件的根元素，这个根元素对所有的 Hibernate 映射文件也都是相同的。<hibernate-mapping.../>元素下面有<class.../>子元素，每个子元素映射一个 POJO 持久化类，在该映射文件里<class name="com.chpt10.model.News" table="news_table">表示这样的含义：name 属性指定对应的持久化类，即 POJO 的名称是 com.chpt10.model.News；table 属性指定该持久化类映射的数据库中的表的名称；当使用<class.../>元素来映射某个持久化类时，通常还需要<id.../>和<property.../>两个最常见的子元素，其中<id.../>元素用于映射标识属性，而<property.../>用于映射普通属性。

　　通常情况下，Hibernate 建议为持久化类定义一个标识属性，用于唯一地标识某个持久化实例，而标识属性则需要映射到底层数据表的主键。标识属性都是通过<id.../>元素来指定的。其中 name 属性为持久化类的标识属性名，type 属性用来指定该标识属性的类型，如果省略则使用与持久化类想对应的属性值；column 属性用来指定标识属性所映射的数据表中数据列的列名，如果省略则数据库表中列的名称与 name 的值相同。在上面的示例中，持久化类 News 的标识属为 id，类型为 int，对应到数据库 news_table 表中的主键是 news_id 列。<generatro.../>用来指定数据库主键的生成策略，代码<generator class="identity" />说明主键的生成器策略是 identigy，这表示 Hibernate 将根据底层数据库的 identity 字段来提供标识属性值，除了 identity 生成器策略，还有其他的主键生成器策略，大家可以去查阅 Hibernate 相关参考书籍。

　　Hibernate 使用<property.../>元素映射持久化类的普通属性到数据库字段的关系。其中 name 属性用来指定持久化类的普通属性的名字，type 用来指定该普通属性的数据类型，<column.../>用来指定数据库表的字段名，如果 name 和 column 的值一样，还可以把 column 省略。

　　通过上面的分析，我们可以这样总结：一个 POJO 对应数据库中的一张表，这样的对应关系式通过 Hibernate 的映射文件来完成，Hibernate 可以让用户通过 POJO 对象来实现对表的相关操作。可以这样说：

对象持久化 ＝ POJO ＋ 映射文件

　　撰写了 POJO 和与之相对应的映射文件，就可以很方便地进行对象的持久化操作了。

10.3.3　修改 Hibernate 配置文件

　　设置完 Hibernate 的映射文件和持久化类过后，还需要修改 Hibernate 的配置文件 hibernate.cfg.xml，告知 Hibernate 增加了一个持久化类到数据库表的映射关系，在 FirstHibernate 项目的 src 目录下，打开 hibernate.cfg.xml 文件，代码如下：

```
<?xml version='1.0' encoding='UTF-8'?>
<!DOCTYPE hibernate-configuration PUBLIC
        "-//Hibernate/Hibernate Configuration DTD 3.0//EN"
        "http://hibernate.sourceforge.net/hibernate-configuration-3.0.dtd">

<!-- Generated by MyEclipse Hibernate Tools.                      -->
```

```xml
<hibernate-configuration>

    <session-factory>
        <property name="hbm2ddl.auto">update</property>
        <property name="dialect">org.hibernate.dialect.MySQLDialect</property>
        <property name="connection.url">jdbc:mysql://localhost:3306/javaee
            </property>
        <property name="connection.username">root</property>
        <property name="connection.password">1234</property>
        <property name="connection.driver_class">com.mysql.jdbc.Driver</property>

            <property name="hibernate.c3p0.max_size">20</property>
            <property name="hibernate.c3p0.min_size">1</property>
            <property name="hibernate.c3p0.timeout">5000</property>
            <property name="hibernate.c3p0.max_statements">100</property>
            <property name="hibernate.c3p0.idle_test_period">3000</property>
            <property name="hibernate.c3p0.acquire_increment">2</property>
            <property name="hibernate.c3p0.validate">true</property>
        <property name="javax.persistence.validation.mode">none</property>

            <!-- Mapping files -->
        <mapping resource="com/chpt10/model/News.hbm.xml"/>

    </session-factory>

</hibernate-configuration>
```

该配置文件为 Hibernate 的默认配置文件，当程序调用 Configuration 对象的 configure 方法时，Hibernate 将自动加载该文件。<mapping.../>元素用来指定映射文件的名字，这样 Hibernate 就会自动加载这个文件，建立持久化类即 POJO 到表之间的映射关系。代码 <mapping resource=" News.hbm.xml"/>表示 Hibernate 将自动加载 News.hbm.xml 文件。以后如需建立多个持久化类到数据表之间的映射，则也需要一一将它们的映射文件添加到此配置文件中。

10.3.4 创建操作数据库的主类：NewsOperator

NewsOperator 是一个 Java 类，它可以使用 Hibernate 提供的接口去操作 POJO 类 News，从而完成对数据库的操作。在数据库中新增记录在 Hibernate 中不需要使用 insert 命令，只需要构造新增的对象后，调用 Session 对象的 save()方法即可。下面是完成消息插入的代码：

```java
package com.chpt10.test;

import org.hibernate.Session;
import org.hibernate.SessionFactory;
import org.hibernate.Transaction;
```

第 10 章 Hibernate 概述及实例分析

```java
import org.hibernate.cfg.Configuration;

import com.chpt10.model.News;

public class NewsOperator {
    public static void main(String[] args) throws Exception {

        Configuration conf = new Configuration().configure();
        SessionFactory sf = conf.buildSessionFactory();
        Session sess = sf.openSession();
        Transaction tx = sess.beginTransaction();
        News news = new News();
        news.setTitle("第一条消息");
        news.setContent("第一条测试消息的内容。。。。。。");
        sess.save(news);
        tx.commit();
        sess.close();
    }
}
```

上面的 NewsOperator 类非常简单，保存消息只需要程序中斜体字所示的代码：首先创建一个 News 对象，再使用 session 的 save()方法保存 News 对象即可，这是纯粹的对象化的操作方式，非常简单方便。当 Java 程序以面向对象的方式来操作持久化对象时，Hibernate 负责将这种操作转换为底层 SQL 操作。

程序中粗体字代码显示：执行 session.save(news)之前，先要获取 Session 对象。持久化类只有在 Session 的管理下才能够完成数据库的访问。为了使用 Hibernate 进行持久化操作，通常的操作步骤如下：

（1）开发持久化类，在编写相应的映射文件；
（2）获取 Configuration 对象；
（3）通过 Configuration 对象获取 SessionFactory 对象；
（4）通过 SessionFactory 对象获取 Session 对象，并打开事务；
（5）用面向对象的方式操作数据库；
（6）关闭事务，关闭 Session。

通过上面的代码，我们知道，对象的持久化操作必须在 Session 的管理下才能够同步到数据库。Session 由 SessionFactory 工厂生成，SessionFactory 是数据库编译后的内存镜像，通常一个应用对应一个 SessionFactory 对象。SessionFactory 对象由 Configuration 对象生成，Configuration 对象负责加载 Hibernate 的配置文件。

下面测试一下 NewsOperator 主类的运行，在 MyEclipse 的左边的导航窗中，右击主程序类 NewsOperator.java，在弹出的快捷菜单中选择"运行方式"→"Java 应用程序"命令，如图 10-22 所示。

随后，通过 Navicat_for_MySQL 打开数据库 javaee，在数据库中多一张表 news_table，双击打开该表，该表中多了一条记录，通过执行一条 save()方法，Hibernate 自动建造了一张表，并成功地向表中添加了一条记录，如图 10-23 所示。

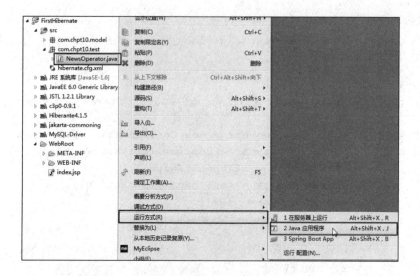

图 10-22 运行项目的主调程序类 NewsOperator

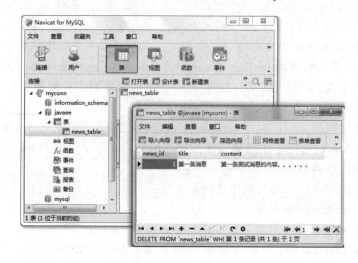

图 10-23 使用 Hibernate 成功插入一条记录

10.3.5 数据查询

Hibernate 提供了非常强大的查询体系,使用 Hibernate 有多种查询方式可以选择,既可以使用 Hibernate 提供的 HQL 进行查询但,也可以使用条件查询,甚至可以使用原来的 SQL 查询语句。在本书中提倡使用 HQL 进行查询操作。HQL 是一种非常强大的语言,这种语言看上去很像 SQL,容易理解和使用。但实际上 HQL 是一种完全面向对象的查询语言,它可以理解如继承、多态和关联等概念。SQL 操作的对象是数据表、列等数据库对象,而 HQL 的操作对象是类、实例、属性等。

HQL 查询依赖 Hibernate 为我们提供的一个 Query 类,每个 Query 实例对应一个查询对象。使用 HQL 语言查询的步骤如下:

(1) 获取 Hibernate Session 对象。
(2) 编写 HQL 语句。

第 10 章　Hibernate 概述及实例分析

（3）以 HQL 语句为参数，调用 Session 的 createQuery 方法创建查询对象。
（4）如果 HQL 包含参数，则调用 Query 的 setXXX()方法为参数赋值。
（5）调用 Query 对象的 list 等方法返回查询结果列表。
下面给出一个使用 HQL 进行查询的示例，代码如下：

```java
package com.chpt10.test;

import java.util.Iterator;
import java.util.List;

import org.hibernate.Session;
import org.hibernate.SessionFactory;
import org.hibernate.Transaction;
import org.hibernate.cfg.Configuration;

import com.chpt10.model.News;

public class NewsQuery {
    public static void main(String[] args) throws Exception {
        Configuration conf = new Configuration().configure();
        SessionFactory sf = conf.buildSessionFactory();
        Session sess = sf.openSession();
        Transaction tx = sess.beginTransaction();

        // 以 HQL 语句创建 Query 对象，执行 setString 方法为 HQL 语句的参数赋值
        // Query 调用 list 方法访问查询的全部实例
        List list = sess
                .createQuery("select n from News n where n.title = :queryTitle")
                .setString("queryTitle", "第一条消息").list();
        // 遍历查询的全部结果
        for(Iterator it = list.iterator(); it.hasNext();) {
            News n = (News) it.next();
            System.out.println(n.getContent());
        }

        tx.commit();
        sess.close();
    }
}
```

由上面的 HQL 语句可以看出，HQL 语句可以使用占位符作为参数。HQL 语句的占位符既可以是英文问号（?，这与 SQL 语句中的占位符一样）；也可以是有名字的占位符。当使用有名字的占位符时，应该在占位符名字的前面增加英文冒号（:），如上面的 HQL 语句。

当进行查询时，编写好 HQL 语句，利用 Session 对象的 createQuery(hql)方法创建一个 Query 对象，该方法的参数即为编写好的 HQL 语句，Query 对象使用 setXXX()方法为

HQL 语句中的参数赋值，如果 HQL 语句中有多个参数，则连续多次调用 setXXX()方法为多个参数赋值，如：

```
sess.createQuery("select n from News n where n.title = :queryTitle and n.content=:queryContent")
    .setString("queryTitle", "第一条消息").setString("","第一条消息的内容...")
```

这得益于 Hibernate 对 Query 类的良好设计。

Query 最后调用 list()方法，将所查询到的符合条件的所有对象放置到一个 List 列表当中返回，通过迭代 List 对象可实现对查询结果的进一步操作。

由于篇幅所限，这里只是简单地介绍了 HQL 的一些用法，关于 HQL 详细的语法及用法介绍，请大家查阅参考相关书籍。

10.3.6 数据编辑

在应用项目开发过程中，数据的编辑或者更新操作是必不可少的一个基本功能，在 Hibernate 框架下的实现非常简单，下面的代码为我们演示了数据更新操作步骤，源程序代码如下：

```java
package com.chpt10.test;

import java.util.Iterator;
import java.util.List;

import org.hibernate.Session;
import org.hibernate.SessionFactory;
import org.hibernate.Transaction;
import org.hibernate.cfg.Configuration;

import com.chpt10.model.News;

public class NewsUpdate {
    public static void main(String[] args) throws Exception {
        Configuration conf = new Configuration().configure();
        SessionFactory sf = conf.buildSessionFactory();
        Session sess = sf.openSession();
        Transaction tx = sess.beginTransaction();

        // 查询出 News 表中所有的记录
        List list = sess.createQuery("select n from News n").list();
        int count = 0;
        // 遍历查询的全部结果
        for(Iterator it = list.iterator(); it.hasNext(); count++) {
            News n = (News) it.next();
            // 更新对象的标题和内容
            n.setTitle("第" + count + "条消息");
            n.setContent("消息" + count + "的内容...");
        }
```

```
            tx.commit();
            sess.close();
    }
}
```

上面的粗体字代码是进行数据更新的关键代码,通过查询将所要更新数据提取到 List 对象当中,通过遍历对象,可以对要修改的对象进行更新即可,不用涉及数据表的任何操作,也不需要保存,当事务提交时,Hibernate 会自动把更新后的信息提交到数据库中,非常方便。

但是这个更新的方式是不提倡的,因为需要先向数据查询出来数据,然后更改后再提交到数据库,而且数据库更新时是逐行更新的,效率十分低下。为了避免这种情况,Hibernate 提供了另外的更新方法,这种方法不仅效率更高,而且代码的书写也更简单,其代码如下:

```
package com.chpt10.test;

import org.hibernate.Session;
import org.hibernate.SessionFactory;
import org.hibernate.Transaction;
import org.hibernate.cfg.Configuration;

public class NewsUpdate2 {
    public static void main(String[] args) throws Exception {

            Configuration conf = new Configuration().configure();
            SessionFactory sf = conf.buildSessionFactory();
            Session sess = sf.openSession();
            Transaction tx = sess.beginTransaction();

            //定义更新的 HQL 语句
            String hqlUpdate = "update News n set n.title = :newTitle";
            //执行更新
            int updatedEntities = sess.createQuery(hqlUpdate)
                        .setString("newTitle", "新消息标题").executeUpdate();

            tx.commit();
            sess.close();
    }
}
```

上面的粗体字代码即为更新语句,可以看出代码更少,效率也更高,而且这种语法也非常类似于 PreparedStatement 的 executeUpdate 语法。同样用过定义不同的 HQL 语句,还可以实现删除,所以在下一小节中的删除操作和此处的代码书写步骤是一样的。

10.3.7 数据删除

在应用项目开发过程中,数据的删除新操作也是必不可少的,在 Hibernate 框架下从

数据库中删除一条记录一般是先取得某对象,然后调用 Session 对象的 delete()方法删除该对象。下面的实例中取得 id(主键)为 3 的 News 对象后,将它删除。下面的代码演示了数据删除操作的步骤,源程序代码如下:

```java
package com.chpt10.test;

import java.util.List;

import org.hibernate.Session;
import org.hibernate.SessionFactory;
import org.hibernate.Transaction;
import org.hibernate.cfg.Configuration;

import com.chpt10.model.News;

public class NewsDelete {
    public static void main(String[] args) throws Exception {
        Configuration conf = new Configuration().configure();
        SessionFactory sf = conf.buildSessionFactory();
        Session sess = sf.openSession();
        Transaction tx = sess.beginTransaction();

        //定义HQL语句
        String hql = "select n from News n where n.id = :queryId";
        //进行查询操作
        List list = sess.createQuery(hql).setInteger("queryId", 3).list();
        //获取要删除的对象,该list列表中只有一个元素
        News news = (News) list.get(0);
        //执行删除
        sess.delete(news);

        tx.commit();
        sess.close();
    }
}
```

如上面的粗体字代码,首先获取要删除的对象,然后通过 Session 对象的 delete()方法即可实现数据表中记录的删除。上面代码中需要注意的一点是,id 为 3 的 News 对象只有一个,但是这一个对象也是放置在 List 列表中,所以要从 list 列表当中获取这个对象,调用 List 对象的 get()方法即可。运行改程序,可以看到数据表 news_table 中 id 为 3 的一条记录已经被删除了。

正如前面的数据更新记录一样,当要批量删除数据的时候,这样的删除方式效率低下,不提倡使用。删除记录推荐使用 HQL 语句实现,这样效率更高。下面使用 HQL 语句实现与上面同样功能的操作,源程序代码如下:

```java
package com.chpt10.test;

import org.hibernate.Session;
import org.hibernate.SessionFactory;
import org.hibernate.Transaction;
import org.hibernate.cfg.Configuration;

public class NewsDelete2 {
    public static void main(String[] args) throws Exception {

        Configuration conf = new Configuration().configure();
        SessionFactory sf = conf.buildSessionFactory();
        Session sess = sf.openSession();
        Transaction tx = sess.beginTransaction();

        //定义删除对象的HQL语句
        String hqlDelete = "delete News n where n.id = :queryId";
        //执行删除操作
        int deleteEntities = sess.createQuery(hqlDelete)
                        .setInteger("queryId", 4).executeUpdate();

        tx.commit();
        sess.close();
    }
}
```

10.3.8 测试

程序编辑完成后要测试一下程序的结果是否正确，其基本的步骤为：在 MyEclipse 的左边的导航窗中，右击要运行的主掉程序类如 NewsOperator.java，在弹出的快捷菜单中选择"运行方式"→"Java 应用程序"命令，如图 10-22 所示。

Hibernate 是一个 ORM 框架，它最终的运行结果都反映在数据库的操作上面，所以须通过 Navicat_for_MySQL 打开数据库 javaee，在数据库中查看运行结果是否正确。比如在执行完 NewsOperator 类后，在数据库中就多一张表 news_table，双击打开该表，该表中多了一条记录。

其他程序的测试运行步骤都与此相同，此处不再赘述。

小　　结

在本章中，我们介绍了使用 Hibernate 框架进行系统开发的通用步骤，围绕 Hibernate 的实用技术详细介绍了 Hibernate 框架的数据增加、数据查询、数据更新编辑以及数据删除等操作的各个方面。而且对相同的数据操作方式提出不同的方法，并对比了不同方法之间的差异。介绍与使用 Hibernate 密切相关的 HQL，通过多个实际的实例程序，演示了

如何灵活地使用 Hibernate 相应的接口和类。通过多个实例观察了在使用和不使用 Hibernate 的情况下，应用程序编程实现的细节差异，详细介绍了如何将 Hibernate 框架集成到自己的应用程序中。

习　题

新建一个工程 Ex10，从网站上下载 Hibernate 的 JAR 包，将其添加到新建的项目工程中，然后在数据库中构建几个表，对该项目增加 Hibernate 支持，配置相应的 Hibernate 文件，让 Hibernate 框架以面向对象的方式操作数据库，实现 hibernate 框架对数据库进行增、删、改和查等操作。在程序编写过程中会出现什么问题？你是如何解决的？

综 合 篇

第 11 章
Spring 入门

学习目标

- 了解 Spring 的作用及意义。
- 掌握 Spring 的安装配置方法。
- 理解和掌握 Spring 配置文件的主要的配置方法。
- 掌握使用 MyEclipse 工具开发 Spring 应用的步骤。

Spring 是一个开源框架，是为了解决企业应用程序开发复杂性而创建的，为 Java EE 应用程序开发提供集成的框架。简单来说，Spring 是一个轻量级的控制反转（IoC）和面向切面（AOP）的容器框架。本章首先介绍了 Spring 框架的基本结构和工作原理，接着介绍了 Spring 框架的下载和安装方法、在 MyEclipse 中增加 Spring 框架支持的详细步骤，通过一个 Spring 实例展示了如何在项目中应用 Spring。通过本章学习，读者能够掌握 Spring 基本编程技术。

11.1 Spring 框架介绍

Spring 从推出至今，一直受到 Java 开发人员的强烈关注。可以说 Spring 是目前最优秀的开源框架之一。Spring 框架由 Rod Johnson 开发的一个开源框架，2003 年发布了 Spring 框架的第一个版本。它是为解决企业应用开发的复杂性而创建的。Spring 使用基本的 JavaBean 来完成以前只可能由 EJB 完成的事情。然而，Spring 的用途不仅限于服务器端的开发。从简单性、可测试性和松耦合的角度而言，任何 Java 应用都可以从 Spring 中受益。

Spring 是一个从实际开发中抽取出来的框架，因此它完成了大量开发中通用步骤，留给开发者的仅仅是与特定应用相关的部分，从而提高了企业应用的开发效率。Spring 提供一种模板式的设计哲学，这些模板完成了大量的重复的步骤，应用这些通用的模板可以节省开发者大量的时间，而把主要的精力用到具体的业务处理上。

Spring 为企业应用开发提供了一种轻量级的解决方案。该方案包括：依赖注入的核心机制，基于 AOP 的声明式事务管理，与多种持久层技术的整合，以及优秀的 Web MVC 框架等。

总结起来，Spring 有如下优点：

（1）低侵入式设计，代码污染极低。

（2）独立于各种应用服务器，可以真正实现"一次编写，处处运行"。

（3）Spring 的 DI 容器降低了业务对象替换的复杂性。

（4）Spring 容器不想取代已有的框架，而是以高度的开发性与它们无缝整合。

（5）Spring 并不完全依赖于 Spring，开发者可自由选用 Spring 框架的部分或全部。

（6）AOP 编程的支持，方便进行面向切面的编程，许多不容易用传统 OOP 实现的功能可以通过 AOP 轻松应付。

图 11-1 显示了 Spring 框架的组成结构。

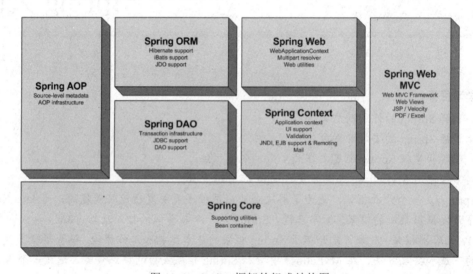

图 11-1 Spring 框架的组成结构图

如图 11-1 所示，当使用 Spring 框架时，必须使用 Spring Core，它代表了 Spring 框架的核心机制，Spring Core 主要由 org.springframework.core、org.springframework.beans 和 org.springframework.context 三个包组成，主要提供 Spring IoC 容器支持。

11.2　Spring 的环境配置

11.2.1　下载 Spring 框架包

笔者成书之时，Spring 的最新的稳定版本是 spring-framework-3.1.2 版本，本书的代码都是基于该版本的 Spring 框架测试通过的。使用 Spring 框架时，需要登录其官方网站 http://www.springsource.org/download，下载 Spring 框架的压缩包 spring-framework-3.1.2.RELEASE.zip 压缩包，该压缩包不仅包含 Spring 的开发包，而且包含 Spring 编译和运行所依赖的第三方类库。为了读者更好地学习 Spring 框架，建议读者使用与本书一致的框架版本。

第 11 章 Spring 入门

解压缩下载后的压缩包，解压缩后得到一个名为 spring-framework-3.1.2.RELEASE 的文件夹，该文件夹下有一个 dist 文件夹，该文件夹下包含了 Spring 框架的所有 JAR 文件，提供了 Spring 的所有功能，但是如果一个 Java EE 项目只是使用到 Spring 的某一方面，通常不需要把所有的 JAR 包都加载进来。下面就经常用到的 JAR 包作简要说明。

（1）org.springframework.aop-3.1.2.RELEASE.jar：Spring 的面向切面编程，提供 AOP（面向切面编程）实现。

（2）org.springframework.asm-3.1.2.RELEASE.jar：Spring 独立的 asm 程序，项目需要额外的 asm.jar 包。

（3）org.springframework.aspects-3.1.2.RELEASE.jar：Spring 提供对 AspectJ 框架的整合。

（4）org.springframework.beans-3.1.2.RELEASE.jar：SpringIoC（依赖注入）的基础实现。

（5）org.springframework.context.support-3.1.2.RELEASE.jar：Spring-context 的扩展支持，用于 MVC 方面。

（6）org.springframework.context-3.1.2.RELEASE.jar：Spring 提供在基础 IoC 功能上的扩展服务，此外还提供许多企业级服务的支持，如邮件服务、任务调度、JNDI 定位、EJB 集成、远程访问、缓存以及各种视图层框架的封装等。

（7）org.springframework.core-3.1.2.RELEASE.jar：Spring 3.1 的核心工具包。

（8）org.springframework.expression-3.1.2.RELEASE.jar：Spring 表达式语言。

（9）org.springframework.instrument.tomcat-3.1.2.RELEASE.jar：Spring 3.1 对 Tomcat 的连接池的集成。

（10）org.springframework.jdbc-3.1.2.RELEASE.jar r：Spring 对 JDBC 的简单封装。

（11）org.springframework.orm-3.1.2.RELEASE.jar：Spring 整合第三方的 ORM 映射支持，如 Hibernate、Ibatis、Jdo 以及 Spring 的 JPA 的支持。

（12）org.springframework.transaction-3.1.2.RELEASE.jar：为 JDBC、Hibernate、JDO、JPA 等提供的一致的声明式和编程式事务管理。

（13）org.springframework.web.servlet-3.1.2.RELEASE.jar：对 Java EE 6.0 Servlet 3.0 的支持。

（14）org.springframework.web.struts-3.1.2.RELEASE.jar：整合 Struts 的支持。

（15）org.springframework.web-3.1.2.RELEASE.jar：SpringWeb 下的工具包。

建议不要将所有的 JAR 文件都复制到应用程序中去，而是当项目需要这个功能或工具时才将相关 JAR 文件加载到应用中去，因为有时候额外多出的 JAR 文件可能引起程序未知的异常。

11.2.2 搭建 Spring 开发环境

开发 Web 项目时，如果使用的开发工具是 MyEclipse，在项目中增加 Spring 框架支持的步骤非常简单。

启动 MyEclipse 工具，在 Web 项目上右击，选择"构建路径"→"添加库"命令，如图 11-2 所示。

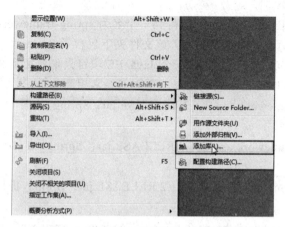

图 11-2　为 Web 项目增加 Spring 类库

出现如图 11-3 所示的"添加库"对话框，选中"用户库"项，然后单击"下一步"按钮，进入选择添加用户库对话框，如图 11-4 所示。

图 11-3　增加用户库对话框　　　　　　图 11-4　选择并添加用户库对话框

MyEclipse 中已经包含了大量的用户库，例如 Hibernate4.1.5 用户库。如果 MyEclipse 中没有用户库，就需要添加自己需要使用的用户库。一个用户库可以包括多个 JAR 文件，添加用户库的作用是使不同的项目可以重复使用。如果是第一次进入该界面，应该是没有任务用户库的。单击"用户库"项，在弹出的首选项对话框中单击"新建"按钮，弹出"新建用户库"对话框，如图 11-5 所示。

在图 11-5 所示的"新建用户库"对话框中输入用户库的名字 spring3.1.2，单击"确定"按钮返回首选项对话框。注意，这里的用户库的名字虽然可以任意命名，但为了用户库的使用者通过名字能清楚地知道其所代表的含义，用户库的命名要使用有一定意义的字符串，如 spring3.1.2 表示该用户库中的 JAR 文件是 Spring 框架 3.1.2 版本的类库文件集合。另外，新建的用户库仅有一个用户库名，还没有包含任何的 JAR 文件，必须向用户库中添加所需的 JAR 文件。

第 11 章　Spring 入门

图 11-5　新建和管理用户库对话框

选中图 11-5 所示对话框中需要编辑的用户库的名字，然后单击"添加外部 JAR"按钮，将出现一个文件浏览对话框，在该文件对话框中选择下载的 Spring 框架解压缩后的 dist 路径下的 org.springframework.core-3.1.2.RELEASE.jar、org.springframework.expression-3.1.2.RELEASE.jar、org.springframework.asm-3.1.2.RELEASE.jar、org.springframework.beans-3.1.2.RELEASE.jar 和 org.springframework.context-3.1.2.RELEASE.jar 文件，单击"打开"按钮，为 spring3.1.2 用户库添加完成 JAR 文件后，看到用户库结构图如图 11-6 所示。

图 11-6　添加完 JAR 文件后 spring3.1.2 用户库的结构图

单击"确定"按钮，返回如图 11-4 所示对话框。在该对话框中勾选 spring3.1.2 和 jakarta-commons 复选框，然后单击如图 11-4 所示对话框的"完成"按钮。为 Web 项目的编译和运行增加用户库成功，MyEclipse 的左边出现如图 11-7 所示的导航树。

至此，项目已经加入 Spring 框架的类库，要在项目中使用 Spring 框架，还需要建立一个 Spring 框架的配置文件 applicationContext.xml，该文件是 Spring 框架的默认配置文

-221-

件，默认放置在项目的 src 目录下。具体操作步骤为：在项目上右击，在弹出的快捷菜单中选择"新建"→"文件"命令，操作过程如图 11-8 所示。

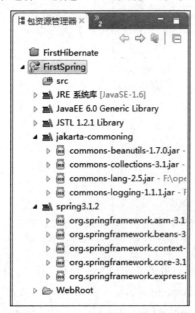
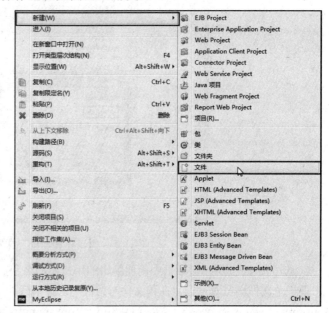

图 11-7　项目成功增加用户库的导航树　　图 11-8　建立 Spring 默认配置文件 applicationContext.xml

之后弹出如图 11-9 所示的"新建文件"对话框，在该对话框中输入存放 Spring 配置文件的父文件夹（配置文件 applicationContext.xml 的存放路径），Spring 配置文件的文件名为 applicationContext.xml。

图 11-9　建立 Spring 默认配置文件 applicationContext.xml 对话框

然后单击"完成"按钮，Spring 配置文件建立成功，该 applicationContext.xml 将被打开，一般的 XML 文件以"设计"视图的形式打开，如图 11-10 所示。

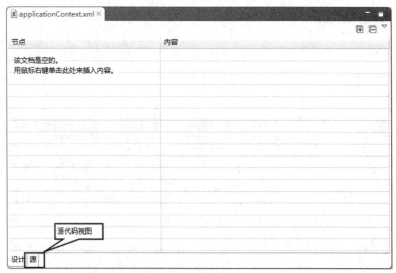

图 11-10 applicationContext.xml 配置文件的配置界面

但这种"设计"可视化视图配置界面的配置操作并不灵活，一般是通常是通过源代码编辑视图界面的方式对 XML 文件进行配置的，单击 XML 文件编辑器的左下角的"源"标签，就切换到了源代码编辑视图界面，之后将下列源代码输入该配置 applicationContext.xml 文件中：

```xml
<?xml version="1.0" encoding="UTF-8"?>
<beans
    xmlns="http://www.springframework.org/schema/beans"
    xmlns:xsi="http://www.w3.org/2001/XMLSchema-instance"
    xmlns:p="http://www.springframework.org/schema/p"
    xsi:schemaLocation="http://www.springframework.org/schema/beans
http://www.springframework.org/schema/beans/spring-beans-3.1.xsd">

</beans>
```

该配置文件的第一行是 XML 文件声明，指定该文件的版本和编码所用的字符集。<beans.../>元素是配置文件的根元素，在该元素内部还可以有多个<bean.../>子元素。

最后，为了方便对日志管理，还需在 Web 项目中增加 log4j.properties 文件，用 log4j 来控制日志输出。log4j.properties 内容文件如下，该文件也放置在 src 路径下。

```
log4j.rootLogger=DEBUG, stdout
log4j.appender.stdout=org.apache.log4j.ConsoleAppender
log4j.appender.stdout.layout=org.apache.log4j.PatternLayout
log4j.appender.stdout.layout.ConversionPattern=%c{1} - %m%n
```

至此，Spring 的开发环境搭建完成，项目开发人员就可以利用 Spring 框架进行应用程序开发，提高项目开发效率。

11.3 第一个 Spring 示例

11.3.1 准备工作

下面开发第一个完整的 Spring 应用程序，通过该应用程序掌握 Spring 的使用方法。首先，启动 MyEclipse 工具，选择 MyEclipse 菜单栏中的"新建"→"Web Project"命令，如图 11-11 所示。

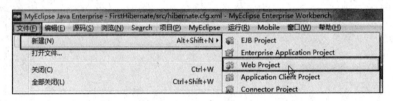

图 11-11 新建一个 Web 项目

弹出新建 Web 项目对话框，将该项目命名为 FirstSpring，选择 JavaEE 版本为 JavaEE 6 – Web 3.0，然后单击"完成"按钮，新建项目成功，如图 11-12 所示。

为新建的 FirstSpring 项目增加 Spring 框架支持，其步骤读者请参看 11.2.2 节所介绍的内容来完成。等这些步骤完成后，在 MyEclipse 左边的导航窗口中出现如图 11-13 所示的导航树。

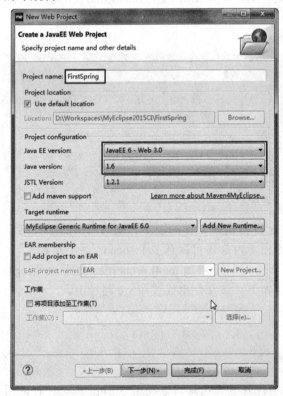

图 11-12 为新建的 Web 项目命名设置　　　　图 11-13 添加 Spring 功能成功

11.3.2 编写接口文件

在开发 Spring 应用程序时，为了充分利用 Spring 框架的控制反转（Inversion of Control，IoC），建议通常面向接口编程，这样也可以更好地让规范和实现分离，从而提供更好的解耦。这样，如果在程序中需要修改或者扩展某个类的功能时，只需要修改或者扩展该接口的实现类，重新修改 Spring 的配置文件即可，而无须修改程序中的其他代码。

在该示例中，首先定义一个 Person 接口，代码如下：

```java
package com.chpt7.component;

public interface Person {
    public void sayHello();
}
```

11.3.3 编写实现接口文件

创建两个不同类，来实现同一接口 Person。一个类为 Chinese，重写接口的 sayHello() 方法，打印字符串为"你好，我来自中国，我的名字是***"；另一个类是 English，也重写接口的 sayHello()方法，打印字符串"Hello, I'm English, my name is ***"。Chinese 类的代码如下：

```java
package com.chpt7.componentImpl;

import com.chpt7.component.Person;

public class Chinese implements Person {
    private String username;
    public String getUsername() {
        return username;
    }
    public void setUsername(String username) {
        this.username = username;
    }
    @Override
    public void sayHello() {
        System.out.println( "你好，我来自中国，我的名字是"+username);
    }
}
```

English 类的代码如下：

```java
package com.chpt7.componentImpl;

import com.chpt7.component.Person;

public class English implements Person {
    private String username;
    public String getUsername() {
        return username;
```

```java
    }
    public void setUsername(String username) {
        this.username = username;
    }
    @Override
    public void sayHello() {
        System.out.println("Hello, I'm English, my name is "+username);
    }
}
```

11.3.4 修改 Spring 的配置文件 applicationContext.xml

为了让 Spring 管理已经开发好的组件，需将其部署到 Spring 的配置文件中，修改 applicationContext.xml 文件，代码如下：

```xml
<?xml version="1.0" encoding="UTF-8"?>
<beans
    xmlns="http://www.springframework.org/schema/beans"
    xmlns:xsi="http://www.w3.org/2001/XMLSchema-instance"
    xmlns:p="http://www.springframework.org/schema/p"
    xsi:schemaLocation="http://www.springframework.org/schema/beans
http://www.springframework.org/schema/beans/spring-beans-3.1.xsd">

    <bean id="person" class="com.chpt7.componentImpl.Chinese">
        <property name="username" value="张三" />
    </bean>

</beans>
```

上面的粗斜体代码部分是在 Spring 配置文件中新增的片段。<beans.../>元素是配置文件的根元素，它可以有多个<bean.../>子元素。每一个<bean>代表一个 Java 组件（Bean 实例），在 Spring 配置文件中配置 Bean 实例通常会指定两个属性：

（1）id：指定该 Bean 的唯一标识，程序通过 id 属性值来访问该 Bean 实例。

（2）class：指定该 Bean 的实现类，此处不能是接口，必须使用实现类，Spring 会使用 XML 解析器读取该属性值，并利用反射来创建该实现类的实例。

Spring 会自动接管每个<bean.../>定义里的<property.../>元素定义，Spring 会在调用无参构造器后、创建默认的 Bean 实例后，调用对应的 setter 方法为程序注入属性值。<property.../>定义的属性值将不再由该 Bean 来主动设置、管理，而是接受 Spring 的注入。

至此，我们定义的 person 这个 Bean 实例的实现类是 Chinese，如果要修改它的实现类为 English，则<bean.../>元素的代码可以修改为：

```xml
<bean id="person" class="com.chpt7.componentImpl.English">
    <property name="username" value="Tom" />
</bean>
```

11.3.5 创建调用组件的主程序类

Spring 帮我们管理配置好的 Bean 实例，如果需要使用这些组件，需编写调用该组件的主程序类，从 Spring 中获取 Bean 实例，并调用 Bean 实例的相关方法。主程序类的代码为：

```java
package com.chpt7.test;
import org.springframework.context.ApplicationContext;
import org.springframework.context.support.ClassPathXmlApplicationContext;
import com.chpt7.component.Person;
public class PersonTest {
    public static void main(String[] args) {
        // 创建Spring容器
        ApplicationContext ctx = new ClassPathXmlApplicationContext(
            "applicationContext.xml");
        // 获取Person实例
        Person p = (Person) ctx.getBean("person");
        // 调用sayHello()方法
        p.sayHello();
    }
}
```

上面程序的两行粗斜体代码实现了创建 Spring 容器，并通过 Spring 容器来获取 id 为 person 的 Bean 实例。一旦通过 Spring 容器获取了 Bean 实例后，使用这个 Bean 实例和一般的 Java 类就没有什么区别了。

从上面的程序可知，id 为 person 的实例的创建不仅不需要了解实例的具体实现，甚至不需要了解实例的创建过程。假如某一天，系统需要改变 person 的实现——这种改变对于实际开发是很常见的，只需要给出 person 的另一个实现，修改 Spring 的配置文件（在本示例中，修改器实现类为 English 即可），而 Person 接口、主程序类的代码无须任何改变。

11.3.6 测试运行

下面演示一下主调程序的运行，其步骤为：

在 MyEclipse 的左边的导航窗中，右击 PersonTest.java，在弹出的快捷菜单中选择"运行方式"→"Java 应用程序"命令，如图 11-14 所示。

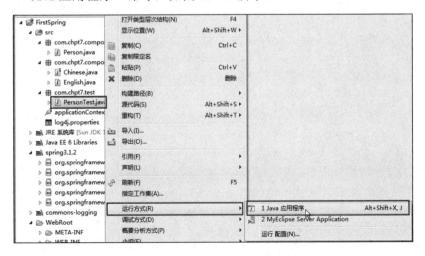

图 11-14 运行项目的主调程序类

随后在 MyEclipse 的控制台视窗中,打印出一条语句:

你好,我来自中国,我的名字是张三

在 Spring 的配置文 applicationContext.xml 中,将 Person 接口的实现类修改为 English,重新运行该主程序类,在 MyEclipse 的控制台视窗中,打印出语句:

Hello, I'm English, my name is Tom

小 结

本章围绕 Spring 的实用技术简要介绍了 Spring 框架的相关方面,包括,Spring 框架的起源、组织结构和工作原理;详细介绍了在如何在实际项目中使用 Spring 框架,以及如何利用 MyEclipse 工具开发 Spring 应用。介绍了 Spring 配置文件的基本结构,以及 Spring 如何通过配置文件自动生成 Bean 实例和管理,倡导面向接口编程,实现组件之间的规范调用和良好解耦;通过一个简单的 Spring 应用实例,让读者深刻理解 Spring 为项目开发带来的便利性。

习 题

新建一个工程 Ex11,从 Spring 的网站上下载 Spring 的 JAR 包,将其添加到新建的项目工程中,然后构建几个存在依赖关系的类,让 Spring 框架通过依赖注入的方式进行管理。在程序编写过程中会出现什么问题?你是如何解决的?

第 12 章
使用 Spring 操作数据库

学习目标
- 掌握 Spring 配置数据源注入的方法。
- 理解 Spring 框架的事务处理原理。
- 理解 PlatformTransactionManager 的接口作用和意义。
- 掌握 Spring 通过 Template 模式访问数据库的方式。

本章更深一步地介绍 Spring 框架的使用和工作模式。包括 Spring 的声明式事务处理模式，Spring 管理普通 Java 实例、Java 组件的方式，Bean 实例的注入方法以及通过 Template 模式访问数据库的访问策略等。Spring 的这些知识都是开发实际应用项目时要用到的，深入理解和掌握这些知识有利于对 Spring 框架的更进一步理解和掌握。

12.1 数据源 datasource 的注入

Spring 框架会对定义在配置文件中的 Bean 实例自动管理，这个 Bean 实例也没有特殊的要求。换句话说，Spring 中的 Bean 就是 Java 实例、Java 组件，它的用法比起传统的 JavaBean 要丰富得多，其功能也要复杂。所以，Spring 框架不仅可以管理标准的 JavaBean，而且可以管理普通的 Java 组件。

在本书的第 10 章，Hibernate 把数据库的连接信息放在了 Hibernate 的配置文件中，为了便于对数据源的管理，可不可以把数据库的连接信息当做一个 Java 组件，配置在 Spring 的配置文件中，让 Spring 框架统一管理呢？答案是肯定的。下面的代码就是将数据库的连接信息配置在了 applicationContext.xml 文件中，让 Spring 框架当做一个普通的 Bean 实例进行管理。

```
<?xml version="1.0" encoding="UTF-8"?>
<beans
    xmlns="http://www.springframework.org/schema/beans"
    xmlns:xsi="http://www.w3.org/2001/XMLSchema-instance"
    xmlns:p="http://www.springframework.org/schema/p"
    xmlns:aop="http://www.springframework.org/schema/aop"
    xmlns:tx="http://www.springframework.org/schema/tx"
    xsi:schemaLocation="http://www.springframework.org/schema/beans
```

```xml
            http://www.springframework.org/schema/beans/spring-beans-3.1.xsd
            http://www.springframework.org/schema/tx
            http://www.springframework.org/schema/tx/spring-tx-3.1.xsd
            http://www.springframework.org/schema/aop
            http://www.springframework.org/schema/aop/spring-aop-3.1.xsd">

    <!-- 定义数据源 Bean，使用 C3P0 数据源实现 -->
    <bean id="dataSource" class="com.mchange.v2.c3p0.ComboPooledDataSource"
        destroy-method="close">
        <!-- 指定连接数据库的驱动 -->
        <property name="driverClass" value="com.mysql.jdbc.Driver"/>
        <!-- 指定连接数据库的 URL -->
        <property name="jdbcUrl" value="jdbc:mysql://localhost/javaee"/>
        <!-- 指定连接数据库的用户名 -->
        <property name="user" value="root"/>
        <!-- 指定连接数据库的密码 -->
        <property name="password" value="1234"/>
        <!-- 指定连接数据库连接池的最大连接数 -->
        <property name="maxPoolSize" value="40"/>
        <!-- 指定连接数据库连接池的最小连接数 -->
        <property name="minPoolSize" value="1"/>
        <!-- 指定连接数据库连接池的初始化连接数 -->
        <property name="initialPoolSize" value="1"/>
        <!-- 指定连接数据库连接池的连接的最大空闲时间 -->
        <property name="maxIdleTime" value="20"/>
    </bean>
</beans>
```

上面的配置文件中，定义了一个 id 为 dataSource 的 Bean 实例，可以通过 Spring 容器获取该实例，进而通过该实例对象轻松获得简单的数据库连接。从该实例也可以看出，Spring 的 Bean 可以代表应用中的任何组件、任何资源实例。

另外，上面配置的<beans...>元素和教材中前面章节的有些不同，这是从 Spring 2.0 开始增加的新特性，使用基于 XML Schema 的配置方式来简化 Spring 的配置文件。上面<beans...>元素中多出的内容是基于 XML Schema 扩展出来的内容。至于 XML Schema 的相关知识此处不做过多介绍，大家可参考相关书籍。

12.2 Spring 框架的事务处理

12.2.1 传统的 JDBC 事务处理

事务是现代数据库理论中的核心概念之一。如果一组处理步骤或者全部发生或者一步也不执行，那么称该组处理步骤为一个事务。当所有的步骤像一个操作一样被完整地执行，那么称该事务被提交。由于其中的一部分或多步执行失败，导致一些步骤没有被提交，则事务必须回滚（回到最初的系统状态）。事务处理在应用程序中起着至关重要的作用，它是一系列任务的组成工作单元，在这个工作单元中，所有的任务要么全部执行,要么全部不执行。以数据库存取的实例来说,就是一组 SQL 指令,这一组 SQL 指令必须全部执行成功,若因为某个原因未全部执行成功(例如其中一行 SQL 有错误),则先前所有执行过的 SQL 指令都会被撤销。

举个简单的例子,一个客户从 A 银行转账至 B 银行,要做的动作是从 A 银行的账户扣款、在 B 银行的账户加上转账的金额,这两个动作必须成功,如果有一个动作失败,则此次转账失败。

在这里介绍 JDBC 如何使用事务管理。首先来看事务的原子性实现。在 JDBC 中,可以操作 Connection 的 setAutoCommit()方法,通过设定 false 参数设置事务的处理方式为人为手动提交,在完成一连串的 SQL 语句后,执行 Connection 的 commit()方法向数据库提交变更,然后恢复 JDBC 事务的默认提交方式,整个事务处理流程执行完成。如果在执行这些一连串的 SQL 语句期间发生错误,则执行 rollback()方法来撤销所有的执行。传统的 JDBC 事务处理的编程步骤一般如下:

```
try {
    .....
    // 更改 JDBC 事务的默认提交方式
    connection.setAutoCommit(false);
    .....
    // 一连串 SQL 操作
    .....
    // 提交 JDBC 事务
    connection.commit();
    // 恢复 JDBC 事务的默认提交方式
    connection.setAutoCommit(true);
} catch(SQLException) {
    // 发生错误,撤销所有变更,回滚 JDBC 事务
    connection.rollback();
}
```

传统的 JDBC 事务处理逻辑要求编程人员每次在进行数据库数据库交互时,都要编写事务处理的代码,这样不但增加了编程人员的工作量,而且还降低了代码的可读性,增加了代码出错的概率。在逻辑关系复杂的项目开发中,这样的事务处理策略是不被推荐使用的。

12.2.2 Spring 框架的事务处理

Spring 框架的成功最吸引人的一点就是 Spring 的事务处理,它提供了一个轻量级的容器事务处理策略。通过 Spring 的事务框架,可以按照统一的编程模型来进行事务编程,却不用关心所使用的数据访问技术以及具体要访问什么类型的事务资源,而且 Spring 的事务框架与 Spring 提供的数据访问支持可以紧密结合,避免了程序员在处理事务的过程与数据访问之间复杂烦琐的工作。

Spring 提供编程式的事务管理(Programmatic transaction manage-ment)与声明式的事务管理(Declarative transaction management)两种不同的事务管理方式。不管哪种事务处理方式,Spring 都供了一致的编程模型。下面了解一下事务管理的这两种方式。

(1)编程式的事务管理:编程式的事务管理可以清楚地控制事务的边界,也就是

让编程人员自行实现事务开始时间、撤销操作的时机、结束时间等，可以实现细粒度的事务控制。

（2）声明式的事务管理：采用声明式的事务管理，事务管理的 API 可以不用介入程序之中，Spring 不须编写任何代码，便可通过实现基于容器的事务管理。从对象的角度来看，它并不知道自己正被纳入事务管理之中，在不需要事务管理的时候，只要在设置文件上修改一下设置，即可移去事务管理服务。

大部分时候推荐使用声明式的事务管理，它允许用户在开发过程中无须理会任何事务逻辑，等到应用开发完以后使用声明式事务进行统一事务管理。只需要在配置文件中增加事务控制片段，业务逻辑组件的方法将会具有事务性。本教材的讲解也是以声明式的事务管理为例进行的。

使用 Spring 声明式事务，Spring 使用 AOP 来支持声明式事务，会根据事务属性，自动在方法调用之前决定是否开启一个事务，并在方法执行之后决定事务提交或回滚事务。从 Spring 2.x 版本以后，使用 XML Schema 方式使事务配置更为简洁明了，Spring 提供了 tx 的命名空间来配置事务管理，tx 命名空间下提供了<tx:advice.../>元素来配置事物切面，一旦使用该元素配置了事务切面，就可直接使用<aop:advisor.../>元素启用自动代理了。

为业务逻辑组件添加事务的步骤如下：

（1）针对不同的事务策略配置对应的事务管理器。

（2）使用<tx:advice.../>元素配置事务切面 Bean，配置事务切面 Bean 时使用多个<method.../>子元素为不同的方法指定相应的事务语义。

（3）在<aop:config.../>元素中使用<aop:advisor.../>元素配置自动事务代理。

假如在 Spring 的配置文件中已经配置好了一个事务管理器（参见 12.3 节的内容），则只需增加如下的代码即可为业务逻辑组件的方法添加事务支持，而再也不用像传统的事务处理那样在代码中书写事务处理的代码。

```xml
<!-- 配置事务切面 Bean,指定事务管理器 -->
<tx:advice id="txAdvice" transaction-manager="transactionManager">
    <!-- 用于配置详细的事务语义 -->
    <tx:attributes>
        <!-- 所有以'get'开头的方法是 read-only 的 -->
        <tx:method name="get*" read-only="true"/>
        <!-- 其他方法使用默认的事务设置 -->
        <tx:method name="*"/>
    </tx:attributes>
</tx:advice>
<aop:config>
    <!-- 配置一个切入点，匹配 lee 包下所有以 Impl 结尾的类执行的所有方法 -->
    <aop:pointcut id="pointcut"
        expression="execution(* com.chpt12.service.impl.*Impl.*(..))"/>
    <!-- 指定在 txAdvice 切入点应用 txAdvice 事务切面 -->
    <aop:advisor advice-ref="txAdvice"
        pointcut-ref="pointcut"/>
</aop:config>
```

如上第一行粗体字代码配置了一个事务切面，配置<tx:advice.../>元素时只需要指定 transaction-manager 属性，该属性的默认值为'transactionManager'。transactionManager 为在 Spring 配置文件中定义的事务管理器 Bean 实例的引用（关于事务管理器的内容参见 12.3 节的相关介绍）。第二行粗体字代码表示的含义为：定义了一个切入点，它匹配 com.chpt12.service.Impl 包下所有以 Impl 结尾的类的所有方法。我们把该切入点称为 pointcut。然后用一个<aop:advisor.../>把这个切入点与 txAdvice 绑定在一起，表示当 com.chpt12.service.Impl 包下所有以 Impl 结尾的类的所有方法执行时，txAdvice 所代表的事务都将执行。

需要注意的一点是，当配置事务切面 Bean 时，要向项目中加载 Spring 提供的 JAR 包为 org.springframework.transaction-3.1.2.RELEASE.jar 和 org.springframework.aop- 3.1.2.RELEASE.jar 文件，只需将这些 JAR 包文件添加到用户库 spring3.1.2 中即可，添加的方法前面章节以后详细介绍，此处不再赘述。

12.3 PlatformTransactionManager 的接口作用

Spring 之所以能够按照一致的编程接口和编程模型来进行事务编程，是通过 PlatformTransactionManager 接口来实现的，该接口是 Spring 事务策略的核心。该接口的源代码如下：

```java
public interface PlatformTransactionManager(){
    //平台无关的获得事务的方法
    TransactionStatue getTransaction(TransactionDefinition definition)
        throws TransactionException;
    //平台无关的事务提交方法
    void commit(TransactionStatus status) throws TransactionException;
    //平台无关的事务返回方法
    void rollback(TransactionStatus status) throws TransactionException;
}
```

PlatformTransactionManager 是一个与任何事务策略分离的接口，随着底层不同事务策略的切换，应用程序须采用不同的实现类，这是 Spring 进行事务管理的一种策略。换句话说，底层的数据库访问技术可以是 JDBC，也可能是 Hibernate，或者是 JDO、JTA 等，PlatformTransactionManager 是所有这些数据库访问技术之上的一个接口，当采用不同的数据库访问技术时只需要指定该接口的一个实现类即可，而对于针对这个接口操作的应用程序无须任何修改，这就实现了编程代码的一致性和良好的解耦。PlatformTransactionManager 代表事务管理的接口，作为编程人员无须知道底层如何管理事务、具体的实现规则是由哪个类来完成的，只要知道事务管理的提交开始事务（getTransaction()）、提交事务（commit()）和回滚事务（rollback()）三个方法就可以了。

从对 PlatformTransactionManager 接口的分析，我们知道 Spring 并不直接管理事务，事实上，它只是提供事务的多方选择，并将事务管理的职责委托给某一个特定的平台实现，比如用 Hibernate 或者是别的持久机制。Spring 提供了许多内置事务管理器实现，它们都实现了 PlatformTransactionManager 接口。

（1）DataSourceTransactionManager：JDBC 事务管理器，提供对单个 javax.sql.DataSource 事务管理，用于 Spring JDBC 抽象框架、iBATIS 或 MyBatis 框架的事务管理。

（2）HibernateTransactionManager：Hibernate 事务管理器，提供对单个 org.hibernate.Session

Factory 事务支持,用于集成 Hibernate 框架时的事务管理;该事务管理器只支持 Hibernate3 以上版本。

（3）JdoTransactionManager：JDO 事务管理器，提供对单个 javax.jdo.PersistenceManagerFactory 事务管理，用于集成 JDO 框架时的事务管理。

（4）JpaTransactionManager：Jpa 事务管理器，提供对单个 javax.persistence.EntityManagerFactory 事务支持，用于集成 JPA 实现框架时的事务管理。

（5）JtaTransactionManager：JTA 事务管理器，提供对分布式事务管理的支持，并将事务管理委托给 Java EE 应用服务器事务管理器。

（6）WebSphereUowTransactionManager：WebSphere 事务管理器，Spring 提供的对 WebSphere 6.0 以上版本应用服务器事务管理器的适配器，此适配器用于对应用服务器提供的高级事务的支持。

有了这些实现类，我们在使用 Spring 的事务抽象框架进行事务管理的时候，只需要根据当前使用的数据访问技术选择对应的 PlatformTransactionManager 实现类即可。

如果应用程序使用 JDBC 进行数据访问，那么可以使用 Spring 提供的 DataSourceTransactionManager 事务管理器，在 Spring 的配置文件中如下配置组件：

```
<!--配置JDBC事务管理器,该类实现PlatformTransactionManager接口-->
<bean id="transactionManager"
class="org.springframework.jdbc.datasource.DataSourceTransactionManager">
    <!-- 配置DataSourceTransactionManager时需要注入DataSource的引用 -->
    <property name="dataSource" ref="dataSource"/>
</bean>
```

在配置该 DataSourceTransactionManager 事务管理器时，需要对该类的 dataSource 属性注入一个 Bean，该 Bean 以引用的形式提供：ref="dataSource"，表明该 dataSource 在 Spring 的配置文件中已经存在，Spring 容器进行实例化的时候，会自动到检测该 dataSource 实例并注入到 DataSourceTransactionManager 事务管理器中。这里用到的 dataSource 实际上就是 12.2 节定义的数据源 dataSource 组件。

需要注意的一点是，当配置 id 为 transactionManager 的组件到 Spring 的配置文件中时，Spring 会提示报错找不到 org.springframework.jdbc.datasource.DataSourceTransactionManager 类。原因是没有加载 Spring 提供的 JAR 包 org.springframework.jdbc-3.1.2.RELEASE.jar 文件，只需将该 JAR 包文件添加到用户库 spring3.1.2 中即可。

如果应用程序使用 Hibernate 持久化进行数据访问，那么可以使用 Spring 提供的 HibernateTransactionManager 事务管理器，在 Spring 的配置文件中如下配置组件：

```
<!-- 配置Hibernate事务管理器,该类实现PlatformTransactionManager接口 -->
<bean id="transactionManager"
    class="org.springframework.orm.hibernate4.HibernateTransactionManager">
    <property name="sessionFactory" ref="sessionFactory" />
</bean>
```

同理，在配置该 DataSourceTransactionManager 事务管理器时，首先要添加 Spring 提供的 JAR 包 org.springframework.orm-3.1.2.RELEASE.jar 文件到用户库 spring3.1.2；然后，需要对该类的 sessionFactory 属性注入一个引用 Bean，该 Bean 是 Hibernate 框架提供的 sessionFactory 类，但是目前 Spring 的配置文件中不存在该组件，如何将 Hibernate 框

架提供的 sessionFactory 对象注入给 Spring 框架中的 Hibernate 事务管理器 HibernateTransaction Manager 的属性？这就涉及 Spring 框架和 Hibernate 框架的整合。

在纯粹的 Hibernate 访问中，应用程序需要手动创建 SessionFactory 实例，这不是以后优秀的策略。在实际开发中，我们希望以一种声明式的方式管理 SessionFactory，直接以配置文件的方式管理 SessionFactory 实例。Spring 框架正好提供了这种管理方式，它不仅能以声明式的方式配置 SessionFactory 实例，还可为 SessionFactory 注入数据源的引用。

下面是在 Spring 的配置文件中配置 SessionFactory 的代码。

```xml
<!-- 定义Hibernate的SessionFactory -->
<bean id="sessionFactory"
    class="org.springframework.orm.hibernate4.LocalSessionFactoryBean">
    <!-- 依赖注入数据源dataSource -->
    <property name="dataSource" ref="dataSource"/>
    <!-- mappingResouces属性用来列出全部映射文件 -->
    <property name="mappingResources">
        <list>
            <!-- 以下用来列出Hibernate映射文件 -->
            <value>News.hbm.xml</value>
        </list>
    </property>
    <!-- 定义Hibernate的SessionFactory的属性 -->
    <property name="hibernateProperties">
        <props>
            <prop key="hibernate.dialect">
                org.hibernate.dialect.MySQLDialect</prop>
            <prop key="hibernate.hbm2ddl.auto">update</prop>
            <prop key="javax.persistence.validation.mode">none</prop>
        </props>
    </property>
</bean>
```

一旦在 Spring 中配置了 SessionFactory 实例，它将随着 Spring 框架的启动而启动，并将该 Bean 实例注入 Hibernate 事务管理器 HibernateTransactionManager 的属性。

从上面的是代码可以看出，Spring 框架代替 Hibernate 管理 SessionFactory 实例，Spring 配置文件中 sessionFactory 实例的定义和 Hibernate 配置文件中 sessionFactory 的配置是一致的，Hibernate 的配置文件 hibernate.cfg.xml 就没有存在的必要了（但是 Hibernate 框架的映射文件是不可缺少的），也就是说 Spring 通过对 sessionFactory 的管理实现了对 Hibernate 框架的整合。

实现 Spring 对 Hibernate 完整的整合，其 applicationContext.xml 文件源代码为：

```xml
<?xml version="1.0" encoding="UTF-8"?>
<beans
    xmlns="http://www.springframework.org/schema/beans"
    xmlns:xsi="http://www.w3.org/2001/XMLSchema-instance"
    xmlns:p="http://www.springframework.org/schema/p"
    xmlns:aop="http://www.springframework.org/schema/aop"
    xmlns:tx="http://www.springframework.org/schema/tx"
    xsi:schemaLocation="http://www.springframework.org/schema/beans
```

```xml
        http://www.springframework.org/schema/beans/spring-beans-3.1.xsd
        http://www.springframework.org/schema/tx
        http://www.springframework.org/schema/tx/spring-tx-3.1.xsd
        http://www.springframework.org/schema/aop
        http://www.springframework.org/schema/aop/spring-aop-3.1.xsd">

    <!-- 定义数据源 Bean，使用 C3P0 数据源实现 -->
    <bean id="dataSource" class="com.mchange.v2.c3p0.ComboPooledDataSource"
        destroy-method="close">
        <!-- 指定连接数据库的驱动 -->
        <property name="driverClass" value="com.mysql.jdbc.Driver"/>
        <!-- 指定连接数据库的 URL -->
        <property name="jdbcUrl" value="jdbc:mysql://localhost/javaee"/>
        <!-- 指定连接数据库的用户名 -->
        <property name="user" value="root"/>
        <!-- 指定连接数据库的密码 -->
        <property name="password" value="1234"/>
        <!-- 指定连接数据库连接池的最大连接数 -->
        <property name="maxPoolSize" value="40"/>
        <!-- 指定连接数据库连接池的最小连接数 -->
        <property name="minPoolSize" value="1"/>
        <!-- 指定连接数据库连接池的初始化连接数 -->
        <property name="initialPoolSize" value="1"/>
        <!-- 指定连接数据库连接池的连接的最大空闲时间 -->
        <property name="maxIdleTime" value="20"/>
    </bean>

    <!-- 定义 Hibernate 的 SessionFactory -->
    <bean id="sessionFactory"
        class="org.springframework.orm.hibernate4.LocalSessionFactoryBean">
        <!-- 依赖注入数据源 dataSource -->
        <property name="dataSource" ref="dataSource"/>
        <!-- mappingResouces 属性用来列出全部映射文件 -->
        <property name="mappingResources">
            <list>
                <!-- 以下用来列出 Hibernate 映射文件 -->
                <value>News.hbm.xml</value>
            </list>
        </property>
        <!-- 定义 Hibernate 的 SessionFactory 的属性 -->
        <property name="nibernateProperties">
            <props>
                <prop key="hibernate.dialect">
                    org.hibernate.dialect.MySQLDialect</prop>
                <prop key="hibernate.hbm2ddl.auto">update</prop>
                <prop key="javax.persistence.validation.mode">none</prop>
            </props>
        </property>
    </bean>
```

第 12 章 使用 Spring 操作数据库

```xml
<!-- 配置Hibernate事务管理器,该类实现PlatformTransactionManager接口 -->
<bean id="transactionManager"
    class="org.springframework.orm.hibernate4.HibernateTransactionManager">
    <property name="sessionFactory" ref="sessionFactory" />
</bean>
<!-- 配置事务切面Bean,指定事务管理器 -->
<tx:advice id="txAdvice" transaction-manager="transactionManager">
    <!-- 用于配置详细的事务语义 -->
    <tx:attributes>
        <!-- 所有以'get'开头的方法是read-only的 -->
        <tx:method name="get*" read-only="true"/>
        <!-- 其他方法使用默认的事务设置 -->
        <tx:method name="*"/>
    </tx:attributes>
</tx:advice>
<aop:config>
    <!-- 配置一个切入点,匹配lee包下所有以Impl结尾的类执行的所有方法 -->
    <aop:pointcut id="pointcut"
        expression="execution(* com.chpt12.service.impl.*Impl.*(..))"/>
    <!-- 指定在txAdvice切入点应用txAdvice事务切面 -->
    <aop:advisor advice-ref="txAdvice"
        pointcut-ref="pointcut"/>
</aop:config>
```

12.4 使用 Template 访问数据

12.4.1 Template 模式简介

Template 模式指的就是在父类中定义一个操作算法的骨架或者说操作顺序，而将一些步骤的具体实现延迟到子类中。父类始终控制着整个流程的主动权，子类只是辅助父类实现某些可定制的步骤。这个模式的运用可以使编程人员的工作变得简单和规范。

Spring 对 JDBC 的抽象和对 Hibernate 的集成，都采用了这样的一种 Template 模式或者处理方式。通过 Template 模式简介与相应的 Callback 接口相结合，以一种统一而集中的方式来处理资源的获取和释放。下面以 HibernateTemplate 为例，看看使用 Template 模式带来的好处。如果应用程序要向数据库中保存一条记录，必须使用如下的代码才能完成：

```
Session session = HibernateUtil.getSession();
    Transaction tx = session.beginTransaction();
    sess.save(news);
    tx.commit();
    HibernateUtil.closeSession();
```

上面的代码还省去了异常处理，同时使用了 HibernateUtil 类来简化从 SessionFactory 获取 Session，以及关闭 Session 等处理。实际上，编程人员真正要做的业务逻辑操作只是上面的粗体字代码的一个保存动作，但是为了保存动作的顺利完成须加上事务处理的很多代码。

HibernateTemplate 是 Spring 提供的模板工具类，通过使用这种简便的模板工具类，可以为我们完成开发中大量需要重复进行的工作，使业务逻辑处理变得十分简单，下面的一行代码即可完成数据的保存操作：

```
getHibernateTemplate().save(user);
```

这种简便的数据操作方法就是 Spring 框架通过 Template 模式提供的。

12.4.2　HibernateTemplate 的使用

HibernateTemplate 提供持久层访问模板化操作，它只需要提供一个 SessionFactory 的引用，就可以执行持久化操作。对于 Web 项目，通常应用启动时会自动创建 ApplicationContext，SessionFactory 和访问数据的组件都处于 Spring 框架的管理之下，因此无须在代码中显示设置，它们的依赖关系通过在 Spring 的配置文件 applicationContext.xml 中来设置。

下面的示例程序是基于 HibernateTemplate 实现的一个数据库访问组件，该组件是一个用于对数据库进行增、删、查、改等操作的数据访问对象（即 Data Access Objects，DAO），所以通常称之为 DAO 组件。

```java
public class NewsDaoImpl extends HibernateDaoSupport implements NewsDao {
    /** 加载 News 实例 */
    public News get(Integer id) {
        return (News) getHibernateTemplate().get(News.class, id);
    }

    /** 保存 News 实例 */
    public Integer save(News news) {
        return (Integer) getHibernateTemplate().save(news);
    }

    /** 修改 News 实例 */
    public void update(News news) {
        getHibernateTemplate().update(news);
    }

    /** 通过 id 属性删除 News 实例 */
    public void delete(Integer id) {
        getHibernateTemplate().delete(get(id));
    }

    /** 通过实例对象删除 News 实例 */
    public void delete(News news) {
        getHibernateTemplate().delete(news);
    }

    /** 根据标题名查找 News */
    public List<News> findByName(String title) {
        return(List<News>) getHibernateTemplate().find(
            "from News n where n.title like ?", title);
```

```
    }
    /** 查询全部 News 实例 */
    public List findAllNews() {
        return (List<News>) getHibernateTemplate().find("from News");
    }
}
```

上面的 DAO 组件基本上完成了对数据库 News 表的大部分操作，而借助于 HibernateTemplate 类对数据库的持久化操作变得非常简单。上面的 DAO 组件继承了 HibernateDaoSupport 类，该类是 Spring 提供的 DAO 支持工具类，它已经完成了大量基础性的工作，对实现 DAO 组件有很大帮助。该类的一个最主要的方法就是 getHibernateTemplate() 方法，调用该方法返回一个 HibernateTemplate 对象。一旦获得了 HibernateTemplate 对象，剩下的 DAO 实现将由 HibernateTemplate 来完成。

另外，该 DAO 组件继承了一个 NewsDao 接口，这也是 Spring 提倡的编程思想，面向接口编程可使程序更好地解耦。NewsDao 接口的代码如下：

```
public interface NewsDao {
    public abstract News get(Integer id);
    public abstract Integer save(News news);
    public abstract void update(News news);
    public abstract void delete(Integer id);
    public abstract void delete(News news);
    public abstract List<News> findByName(String title);
    public abstract List findAllNews();
}
```

大部分情况下，通过 HibernateTemplate 这些实用方法就可以完成大多数 DAO 对象的增、删、查、改等操作，下面是 HibernateTemplate 的常用方法的简介。

- void delete(Object entity)：删除指定持久化实例。
- deleteAll(Collection entities)：删除集合内全部持久化类实例。
- find(String queryString)：根据 HQL 查询字符串来返回实例集合。
- findByNamedQuery(String queryName)：根据命名查询返回实例集合。
- get(Class entityClass, Serializable id)：根据主键加载特定持久化类的实例。
- save(Object entity)：保存新的实例。
- saveOrUpdate(Object entity)：根据实例状态，选择保存或者更新。
- update(Object entity)：更新实例的状态，要求 entity 是持久状态。
- setMaxResults(int maxResults)：设置分页的大小。

查看上面的 DAO 组件，不难看出 HibernateTemplate 持久化访问的简洁性，大部分数据库访问操作只需要一行代码即可，完全避免了 Hibernate 持久化操作那些烦琐的步骤。

需要注意的是，该 DAO 组件须纳入 Spring 框架的管理，也就是说必须在 Spring 配置文件中配置一个 NewsDaoImpl，且在定义该 Bean 时需要传入一个 SessionFactory 对象，利用 Spring 框架实现依赖注入，配置文件的代码为：

```
<?xml version="1.0" encoding="UTF-8"?>
<beans
```

```xml
xmlns="http://www.springframework.org/schema/beans"
xmlns:xsi="http://www.w3.org/2001/XMLSchema-instance"
xmlns:p="http://www.springframework.org/schema/p"
xmlns:aop="http://www.springframework.org/schema/aop"
xmlns:tx="http://www.springframework.org/schema/tx"
xsi:schemaLocation="http://www.springframework.org/schema/beans
http://www.springframework.org/schema/beans/spring-beans-3.1.xsd
http://www.springframework.org/schema/tx
http://www.springframework.org/schema/tx/spring-tx-3.1.xsd
http://www.springframework.org/schema/aop
http://www.springframework.org/schema/aop/spring-aop-3.1.xsd">

<!-- 定义数据源 Bean，使用 C3P0 数据源实现 -->
<bean id="dataSource" class="com.mchange.v2.c3p0.ComboPooledDataSource"
    destroy-method="close">
    <!-- 指定连接数据库的驱动 -->
    <property name="driverClass" value="com.mysql.jdbc.Driver"/>
    <!-- 指定连接数据库的 URL -->
    <property name="jdbcUrl" value="jdbc:mysql://localhost/javaee"/>
    <!-- 指定连接数据库的用户名 -->
    <property name="user" value="root"/>
    <!-- 指定连接数据库的密码 -->
    <property name="password" value="1234"/>
    <!-- 指定连接数据库连接池的最大连接数 -->
    <property name="maxPoolSize" value="40"/>
    <!-- 指定连接数据库连接池的最小连接数 -->
    <property name="minPoolSize" value="1"/>
    <!-- 指定连接数据库连接池的初始化连接数 -->
    <property name="initialPoolSize" value="1"/>
    <!-- 指定连接数据库连接池的连接的最大空闲时间 -->
    <property name="maxIdleTime" value="20"/>
</bean>

<!-- 定义 Hibernate 的 SessionFactory -->
<bean id="sessionFactory"
    class="org.springframework.orm.hibernate4.LocalSessionFactoryBean">
    <!-- 依赖注入数据源 dataSource -->
    <property name="dataSource" ref="dataSource"/>
    <!-- mappingResouces 属性用来列出全部映射文件 -->
    <property name="mappingResources">
        <list>
            <!-- 以下用来列出 Hibernate 映射文件 -->
            <value>News.hbm.xml</value>
        </list>
    </property>
    <!-- 定义 Hibernate 的 SessionFactory 的属性 -->
```

第 12 章 使用 Spring 操作数据库

```xml
            <property name="hibernateProperties">
                <props>
                    <prop key="hibernate.dialect">
                        org.hibernate.dialect.MySQLDialect</prop>
                    <prop key="hibernate.hbm2ddl.auto">update</prop>
                    <prop key="javax.persistence.validation.mode">none</prop>
                </props>
            </property>
        </bean>

        <!-- 配置 Hibernate 事务管理器,该类实现 PlatformTransactionManager 接口 -->
        <bean id="transactionManager"
            class="org.springframework.orm.hibernate4.HibernateTransactionManager">
            <property name="sessionFactory" ref="sessionFactory" />
        </bean>
        <!-- 配置事务切面 Bean,指定事务管理器 -->
        <tx:advice id="txAdvice" transaction-manager="transactionManager">
            <!-- 用于配置详细的事务语义 -->
            <tx:attributes>
                <!-- 所有以'get'开头的方法是 read-only 的 -->
                <tx:method name="get*" read-only="true"/>
                <!-- 其他方法使用默认的事务设置 -->
                <tx:method name="*"/>
            </tx:attributes>
        </tx:advice>
        <aop:config>
            <!-- 配置一个切入点,匹配 lee 包下所有以 Impl 结尾的类执行的所有方法 -->
            <aop:pointcut id="pointcut"
                expression="execution(* com.chpt12.service.impl.*Impl.*(..))"/>
            <!-- 指定在 txAdvice 切入点应用 txAdvice 事务切面 -->
            <aop:advisor advice-ref="txAdvice"
                pointcut-ref="pointcut"/>
        </aop:config>
        <!-- 定义 DAO Bean-->
        <bean id="newsDao" class="com.chpt12.dao.impl.NewsDaoImpl">
            <!-- 注入持久化操作所需的 SessionFactory -->
            <property name="sessionFactory" ref="sessionFactory"/>
        </bean>
```

通过上面的配置,已经完成了 Spring 对 Hibernate 的整合,并充分利用 Spring 提供的工具类非常方便地对数据库行了数据持久化操作。

小　　结

本章围绕 Spring 的数据访问技术介绍了 Spring 框架事务处方面的知识,包括 Spring

框架事务处理的思想原理、事务处理的方式、事务处理的设计模式,以及与传统事务处理方式的对比;着重介绍了声明式事务处理的方法和详尽的配置方法;详细介绍了 Spring 对 Hibernate 框架的整合支持和步骤,Spring 的 Template 模式的原理和以及在整合 Hibernate 时的应用。Spring 提供的工具类 HibernateTemplate 和 HibernateDaoSupport 能够简化数据库操作,通过这工具类的灵活运用设计了一个访问数据库操作的 DAO 组件,并给出了该组件访问数据的步骤。

习 题

新建一个工程 Ex12,在本工程中实现三个框架的整合。在整合过程中会出现什么问题?你是怎样解决的?

第 13 章 Spring+Struts2+Hibernate 集成实例

学习目标

- 掌握 Spring+Struts2+Hibernate 集成的方法与步骤。
- 理解系统需求分析的基本思路。
- 理解 Java EE 应用分层模型与总体架构方案。
- 掌握数据库访问组件的实现方式。
- 掌握业务控制器的作用和创建方法。
- 掌握 Web 层视图组件的创建。

本章详细介绍了 Spring、Struts2 和 Hibernate 这三个框架的整合原理和步骤，并完整地介绍了一个 Java EE 项目：网上书店系统，在此系统的基础上可扩展出网络商城、在线电子商务系统等。本章的内容是对前面介绍的三个框架相关知识点和内容的回顾和复习，也是将理论知识运用到实际开发中的实践和尝试，一旦读者掌握了本章案例的开发思路，会对 Spring、Struts2 和 Hibernate 这三个框架有更深一层的理解，深入领会和掌握这些知识，也会对实际 Java EE 项目开发有更高层次的认识和掌握。

13.1 项目需求

13.1.1 项目需求概述

网上书店系统也称购物系统、网上商城系统，是方便企业及个人商家在网上产品展示及在线购物电子商务的系统。企业及个人商家通过它可以建立自己的网上商店，建立网上销售渠道，可以让企业直接面对最终用户，减少销售过程中的中间环节，降低客户的购买成本。使用该系统，用户可以让所有上网浏览的客户看到在网络上所登录的产品信息，并可以查看、购买商品和下订单，缩短与小客户的距离，直接获得效益。

网上书店系统要达到的目标：

（1）用户注册后，登录到网上购物系统中，可以进入购物流程（或在结账之前注册并登录系统）。

（2）客户可以登录网上购物系统浏览和购买图书，当注册后，可以在客户所购图书总金额达一定数量时，从普通会员根据量值不同自动升级成为不同等级的 VIP 会员，并享受不同折扣优惠。

（3）客户一次可以购买多种图书，当确认购买之后，将产生一张订单。

（4）订单生成后，客户可在前台查询订单的处理状态："T"表示订单上的商品已发出，"F"表示订单上的商品未发出。若订单不能在规定日期送达，客户可退单或做其他处理。

（5）客户和商家可在系统发表留言或评论。

本书的网上书店系统仅是从 Java EE 项目开发的角度进行三个框架的综合应用，帮助大家灵活使用这些框架的能力，侧重点在于大家对项目开发知识的理解和掌握，只完成了该网上书店系统的部分功能，其系统功能的完备性和实用性还无法满足真正的项目需求。从此目的出发，将网上书店系统的功能需求定位为：

1．客户前台功能：

（1）客户管理功能。系统实行会员注册或登录，对客户的相关信息的信息（包括用户信息和密码）可以进行修改。

（2）商品信息分类查看功能。根据商品的不同种类（例如畅销图书、推荐图书、最新上架图书、特价图书等不同类型）进行信息的分类查看功能。

（3）商品信息查询功能。为客户提供查询商品信息、搜寻商品的功能，包括按照书名进行模糊查询、按照出版社查询、按照作者查询、按照图书类别进行查询等。

（4）购买功能。客户确定购买对象、下订单、进入购物车界面，完成图书购买功能。在购物车界面还可以自动增、删、改图书商品的种类和数量，图书总金额的自动计算等功能。

（5）订单查看管理功能。用户可以查看自己购买商品的订单信息，可以对订单的清单内容查看和管理等。

2．商家后台功能

（1）图书商品信息管理功能。由系统管理员管理整个系统的商品信息，发布销售商品信息，在系统后台随时增添、修改、更新销售商品信息。

（2）商品信息查询功能。为客户提供查询商品信息、搜寻商品的功能，包括按照书名进行模糊查询、按照出版社查询、按照作者查询、按照图书类别进行查询等。

（3）订单管理功能。系统内构成由订单生成，以及相应的订单管理功能。

13.1.2 系统框架

本章将 Spring + Struts2+Hibernate 进行整合开发 Java Web 项目。网上书店系统结构图，如图 13-1 所示。

第 13 章　Spring+Struts2+Hibernate 集成实例

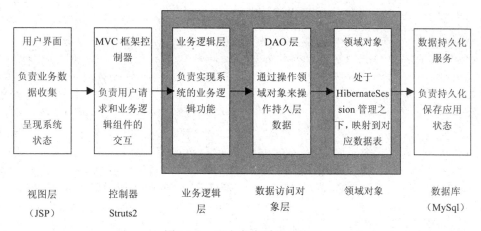

图 13-1　网上书店系统结构图

本系统使用 Spring 框架作为 Hibernate、Struts2 框架的控制器的容器，实现了 Spring、Struts2 和 Hibernate 的集成。

13.2　数据库的设计

根据需求设计如下数据表结构，下面直接给出 SQL 脚本，代码如下：

```sql
-- ----------------------------
-- Table structure for 'bargain'
-- ----------------------------
DROP TABLE IF EXISTS 'bargain';
CREATE TABLE 'bargain' (
  'bargainId' int(11) NOT NULL AUTO_INCREMENT,
  'bookId' int(11) NOT NULL,
  'bookNewPrice' double(10,2) NOT NULL,
  PRIMARY KEY ('bargainId')
) ENGINE=InnoDB AUTO_INCREMENT=4 DEFAULT CHARSET=gb2312;

-- ----------------------------
-- Records of bargain
-- ----------------------------
INSERT INTO 'bargain' VALUES ('1', '1', '10.00');
INSERT INTO 'bargain' VALUES ('2', '6', '12.00');
INSERT INTO 'bargain' VALUES ('3', '5', '11.00');

-- ----------------------------
-- Table structure for 'book'
-- ----------------------------
DROP TABLE IF EXISTS 'book';
CREATE TABLE 'book' (
  'bookId' int(11) NOT NULL AUTO_INCREMENT,
  'bookNumber' varchar(21) DEFAULT NULL,
  'bookName' varchar(20) NOT NULL,
```

```sql
    'bookAuthor' varchar(20) NOT NULL,
    'bookPress' varchar(20) NOT NULL,
    'bookPicture' varchar(100) NOT NULL,
    'bookAmount' int(11) NOT NULL,
    'typeId' int(11) NOT NULL,
    'bookShelveTime' timestamp NOT NULL DEFAULT CURRENT_TIMESTAMP ON UPDATE CURRENT_TIMESTAMP,
    'bookPrice' double(10,2) NOT NULL,
    'bookRemark' varchar(200) DEFAULT NULL,
    'bookSales' int(11) NOT NULL,
    PRIMARY KEY ('bookId')
) ENGINE=InnoDB AUTO_INCREMENT=17 DEFAULT CHARSET=gb2312;

-- ----------------------------
-- Records of book
-- ----------------------------
INSERT INTO 'book' VALUES ('1', '123456789123456778', '红楼梦', '曹雪芹', 'XXX出版社', 'hlm001.jpg', '123', '1', '2012-03-25 14:42:22', '15.00', '披阅十年,增删五次', '2');
INSERT INTO 'book' VALUES ('5', '12345612345612345', '西游记', '吴承恩', 'XXX出版社', 'xyj001.jpg', '32', '1', '2012-03-05 15:09:00', '12.00', '漫漫取经路', '3');
INSERT INTO 'book' VALUES ('6', '12345612345612345', '水浒传', '施耐庵', 'XXX出版社', 'shz001.jpg', '325', '1', '2012-03-05 15:16:24', '23.00', '替天行道', '1');
INSERT INTO 'book' VALUES ('7', '12345612345612345', '三国演义', '罗贯中', 'XXX出版社', 'sgyy001.jpg', '666', '1', '2012-03-05 15:24:05', '20.00', '天下三分', '5');
INSERT INTO 'book' VALUES ('9', '12345612345612345', '红楼梦之史湘云', '小起', '电子科技大学出版社', 'hlm003.jpg', '65', '1', '2012-11-03 21:54:34', '30.00', '其实没有这本书', '0');
INSERT INTO 'book' VALUES ('12', '12345612345612345', '史记', '司马迁', 'XXX出版社', 'sj.jpg', '123', '2', '2012-03-16 15:36:05', '23.00', '史家之绝唱,无韵之离骚', '0');
INSERT INTO 'book' VALUES ('13', 'TSBH090317094422379373', '大秦帝国', '孙皓晖', 'XXX出版社', 'dqdg.jpg', '654', '1', '2012-03-01 09:42:23', '56.00', '纠纠老秦,共赴国难', '0');
INSERT INTO 'book' VALUES ('14', 'TSBH121105000030827754', '快乐密码/心理测试1000问', '(英)托马斯', 'XXX出版社', '12110500030898302.jpg', '10', '5', '2012-11-05 00:03:08', '10.00', '自己是最好的心理医生。通过阅读本书,你可以更好地了解自己,帮助自己', '0');
INSERT INTO 'book' VALUES ('16', 'TSBH121106102272697597', '红高粱家族(新版)', '莫言', '上海文艺出版社', '12111110503331871.jpg', '23', '5', '2012-11-06 10:27:26', '23.00', '《红高粱家族》是莫言1986年向汉语文学、乃至世界文学奉献的一部影响巨大的作品,被译为二十余种文字在全世界发行。', '0');

-- ----------------------------
-- Table structure for 'manager'
-- ----------------------------
DROP TABLE IF EXISTS 'manager';
```

```sql
CREATE TABLE 'manager' (
  'managerId' int(11) NOT NULL AUTO_INCREMENT,
  'managerName' varchar(16) NOT NULL,
  'managerPassword' varchar(12) NOT NULL,
  PRIMARY KEY ('managerId')
) ENGINE=InnoDB AUTO_INCREMENT=2 DEFAULT CHARSET=gb2312;

-- ----------------------------
-- Records of manager
-- ----------------------------
INSERT INTO 'manager' VALUES ('1', 'admin', 'admin');

-- ----------------------------
-- Table structure for 'orders'
-- ----------------------------
DROP TABLE IF EXISTS 'orders';
CREATE TABLE 'orders' (
  'ordersId' int(11) NOT NULL AUTO_INCREMENT,
  'userId' int(11) NOT NULL,
  'ordersTime' timestamp NOT NULL DEFAULT CURRENT_TIMESTAMP ON UPDATE CURRENT_TIMESTAMP,
  'isDeal' char(1) NOT NULL,
  'ordersNumber' varchar(21) DEFAULT NULL,
  'totalMoney' double(10,2) NOT NULL,
  PRIMARY KEY ('ordersId')
) ENGINE=InnoDB AUTO_INCREMENT=18 DEFAULT CHARSET=gb2312;

-- ----------------------------
-- Records of orders
-- ----------------------------
INSERT INTO 'orders' VALUES ('5', '1', '2012-03-16 11:01:53', '1', 'DDBH090031310105125045', '21.00');
INSERT INTO 'orders' VALUES ('6', '1', '2012-03-16 11:01:53', '0', 'DDBH090031316034936676', '20.00');
INSERT INTO 'orders' VALUES ('9', '1', '2012-03-16 11:01:53', '0', 'DDBH090031316042681842', '53.00');
INSERT INTO 'orders' VALUES ('10', '1', '2012-03-16 11:01:53', '1', 'DDBH090031316043280000', '73.00');
INSERT INTO 'orders' VALUES ('11', '1', '2012-03-16 11:01:53', '1', 'DDBH090031316045277898', '84.00');
INSERT INTO 'orders' VALUES ('12', '1', '2012-03-16 11:01:53', '0', 'DDBH090031316061638196', '104.00');
INSERT INTO 'orders' VALUES ('13', '1', '2012-03-16 11:01:53', '0', 'DDBH090031316064167108', '20.00');
INSERT INTO 'orders' VALUES ('14', '1', '2012-03-13 16:06:46', '1', 'DDBH090031316064632612', '20.00');
INSERT INTO 'orders' VALUES ('15', '1', '2012-03-13 16:06:51', '0', 'DDBH090031316065131785', '11.00');
INSERT INTO 'orders' VALUES ('16', '1', '2012-03-13 16:06:58', '0', 'DDBH090031316065881351', '10.00');
```

```sql
INSERT INTO 'orders' VALUES ('17', '1', '2012-03-16 17:19:49', '1', 'DDBH09031617194991827', '83.00');

-- ----------------------------
-- Table structure for 'ordersbook'
-- ----------------------------
DROP TABLE IF EXISTS 'ordersbook';
CREATE TABLE 'ordersbook' (
  'ordersBookId' int(11) NOT NULL AUTO_INCREMENT,
  'ordersId' int(11) NOT NULL,
  'bookId' int(11) NOT NULL,
  'bookAmount' int(11) NOT NULL,
  PRIMARY KEY ('ordersBookId')
) ENGINE=InnoDB AUTO_INCREMENT=28 DEFAULT CHARSET=gb2312;

-- ----------------------------
-- Records of ordersbook
-- ----------------------------
INSERT INTO 'ordersbook' VALUES ('5', '2', '12', '1');
INSERT INTO 'ordersbook' VALUES ('6', '2', '7', '1');
INSERT INTO 'ordersbook' VALUES ('7', '3', '7', '1');
INSERT INTO 'ordersbook' VALUES ('8', '3', '5', '1');
INSERT INTO 'ordersbook' VALUES ('9', '4', '1', '1');
INSERT INTO 'ordersbook' VALUES ('10', '5', '5', '1');
INSERT INTO 'ordersbook' VALUES ('11', '5', '1', '1');
INSERT INTO 'ordersbook' VALUES ('12', '6', '7', '1');
INSERT INTO 'ordersbook' VALUES ('13', '7', '5', '1');
INSERT INTO 'ordersbook' VALUES ('14', '8', '1', '1');
INSERT INTO 'ordersbook' VALUES ('15', '9', '6', '1');
INSERT INTO 'ordersbook' VALUES ('16', '10', '7', '1');
INSERT INTO 'ordersbook' VALUES ('17', '11', '5', '1');
INSERT INTO 'ordersbook' VALUES ('18', '12', '7', '1');
INSERT INTO 'ordersbook' VALUES ('19', '13', '7', '1');
INSERT INTO 'ordersbook' VALUES ('20', '14', '7', '1');
INSERT INTO 'ordersbook' VALUES ('21', '15', '5', '1');
INSERT INTO 'ordersbook' VALUES ('22', '16', '1', '1');
INSERT INTO 'ordersbook' VALUES ('23', '17', '1', '1');
INSERT INTO 'ordersbook' VALUES ('24', '17', '5', '1');
INSERT INTO 'ordersbook' VALUES ('25', '17', '7', '1');
INSERT INTO 'ordersbook' VALUES ('26', '17', '6', '1');
INSERT INTO 'ordersbook' VALUES ('27', '17', '9', '1');

-- ----------------------------
-- Table structure for 'recommended'
-- ----------------------------
DROP TABLE IF EXISTS 'recommended';
CREATE TABLE 'recommended' (
  'recommendedId' int(11) NOT NULL AUTO_INCREMENT,
  'bookId' int(11) NOT NULL,
  PRIMARY KEY ('recommendedId')
```

```sql
) ENGINE=InnoDB AUTO_INCREMENT=4 DEFAULT CHARSET=gb2312;

-- ----------------------------
-- Records of recommended
-- ----------------------------
INSERT INTO 'recommended' VALUES ('1', '1');
INSERT INTO 'recommended' VALUES ('2', '7');
INSERT INTO 'recommended' VALUES ('3', '5');

-- ----------------------------
-- Table structure for 'sex'
-- ----------------------------
DROP TABLE IF EXISTS 'sex';
CREATE TABLE 'sex' (
  'sexId' int(11) NOT NULL AUTO_INCREMENT,
  'sexType' varchar(4) NOT NULL,
  PRIMARY KEY ('sexId')
) ENGINE=InnoDB AUTO_INCREMENT=4 DEFAULT CHARSET=gb2312;

-- ----------------------------
-- Records of sex
-- ----------------------------
INSERT INTO 'sex' VALUES ('1', '男');
INSERT INTO 'sex' VALUES ('2', '女');
INSERT INTO 'sex' VALUES ('3', '未知');

-- ----------------------------
-- Table structure for 'type'
-- ----------------------------
DROP TABLE IF EXISTS 'type';
CREATE TABLE 'type' (
  'typeId' int(11) NOT NULL AUTO_INCREMENT,
  'typeName' varchar(16) NOT NULL,
  PRIMARY KEY ('typeId')
) ENGINE=InnoDB AUTO_INCREMENT=6 DEFAULT CHARSET=gb2312;

-- ----------------------------
-- Records of type
-- ----------------------------
INSERT INTO 'type' VALUES ('1', '文学');
INSERT INTO 'type' VALUES ('2', '历史');
INSERT INTO 'type' VALUES ('3', '天文');
INSERT INTO 'type' VALUES ('4', '地理');
INSERT INTO 'type' VALUES ('5', '其他');

-- ----------------------------
-- Table structure for 'user'
-- ----------------------------
DROP TABLE IF EXISTS 'user';
CREATE TABLE 'user' (
```

```sql
  'userId' int(11) NOT NULL AUTO_INCREMENT,
  'userName' varchar(16) NOT NULL,
  'userPassword' varchar(12) NOT NULL,
  'userEmail' varchar(100) NOT NULL,
  'userNickname' varchar(10) DEFAULT NULL,
  'sexId' int(11) NOT NULL,
  'userAddress' varchar(200) DEFAULT NULL,
  'userPhone' varchar(24) DEFAULT NULL,
  'userRemark' varchar(200) DEFAULT NULL,
  PRIMARY KEY ('userId')
) ENGINE=InnoDB AUTO_INCREMENT=6 DEFAULT CHARSET=gb2312;

-- ----------------------------
-- Records of user
-- ----------------------------
INSERT INTO 'user' VALUES ('1', 'xiaoqi', 'xiaoqi', 'xiaoqi@163.com.cn', '小起', '1', 'fdsfdfdfdf', '123456789123', 'adasdasdasd');
INSERT INTO 'user' VALUES ('2', 'candy', 'candy0101', 'candy@163.com', null, '3', null, null, null);
INSERT INTO 'user' VALUES ('4', 'zhangsan', '123456', 'dsfsf@163.com', null, '3', null, null, null);
INSERT INTO 'user' VALUES ('5', 'weisss', '1234567', 'zhangfeif@163.com', '张飞', '1', '', '12345678', '无');

-- ----------------------------
-- View structure for '123'
-- ----------------------------
DROP VIEW IF EXISTS '123';
CREATE ALGORITHM=UNDEFINED DEFINER='root'@'localhost' SQL SECURITY DEFINER VIEW '123' AS select 'book'.'bookId' AS 'bookId','book'.'bookNumber' AS 'bookNumber','book'.'bookName' AS 'bookName','book'.'bookAuthor' AS 'bookAuthor','book'.'bookPress' AS 'bookPress','book'.'bookPicture' AS 'bookPicture','book'.'bookAmount' AS 'bookAmount','book'.'typeId' AS 'typeId','book'.'bookShelveTime' AS 'bookShelveTime','book'.'bookPrice' AS 'bookPrice','book'.'bookRemark' AS 'bookRemark','book'.'bookSales' AS 'bookSales','bargain'.'bookNewPrice' AS 'bookNewPrice','bargain'.'bargainId' AS 'bargainId' from ('bargain' join 'book') ;
```

数据库中共有9张表，它们所存储的信息分别为：

book：存储图书的信息。

type：图书类别信息。

bargain：特价图书信息。

recommended：推荐图书信息。

orders：订单信息。

ordersbook：订单内容信息，即表示某订单中有多少件图书。

manager：管理员用户信息。

user：普通用户信息。

sex：性别信息。

第 13 章 Spring+Struts2+Hibernate 集成实例

13.3 配置开发环境

启动 MyEclipse，创建一个 Web 工程，命名为 ShoppingOnline。本项目采用 Spring + Struts2+Hibernate 的技术架构进行项目开发，所以这就涉及三个开源框架的整合。Spring 和 Hibernate 的整合在第 12 章已经详细介绍过，剩下的是 Spring 框架和 Struts2 框架的整合。前面介绍的 Spring 容器都是通过手工的方式进行创建的，但对于使用 Spring 的 Web 应用程序，可以通过配置文件声明式地创建 Spring 容器。为了让 Spring 容器随着 Web 应用的启动而启动，可以直接在 web.xml 文件中配置创建 Spring 容器。

13.3.1 web.xml 文件的配置

配置 web.xml 文件，内容如下：

```xml
<?xml version="1.0" encoding="UTF-8"?>
<web-app xmlns="http://java.sun.com/xml/ns/javaee"
    xmlns:xsi="http://www.w3.org/2001/XMLSchema-instance"
    xsi:schemaLocation="http://java.sun.com/xml/ns/javaee
    http://java.sun.com/xml/ns/javaee/web-app_2_5.xsd" version="2.5">
    <!-- 配置Spring配置文件的位置 -->
    <context-param>
        <param-name>contextConfigLocation</param-name>
        <param-value>/WEB-INF/applicationContext.xml</param-value>
    </context-param>
    <!-- 使用ContextLoaderListener初始化Spring容器 -->
    <listener>
<listener-class>org.springframework.web.context.ContextLoaderListener
        </listener-class>
    </listener>
    <!-- 定义Struts2的FilterDispathcer的Filter -->
    <filter>
        <filter-name>struts2</filter-name>
        <filter-class>org.apache.struts2.dispatcher.FilterDispatcher
        </filter-class>
    </filter>
    <!-- FilterDispatcher用来初始化Struts2并且处理所有的WEB请求。 -->
    <filter-mapping>
        <filter-name>struts2</filter-name>
        <url-pattern>/*</url-pattern>
    </filter-mapping>

    <!-- 定义Web应用的首页 -->
    <welcome-file-list>
        <welcome-file>index.jsp</welcome-file>
    </welcome-file-list>
</web-app>
```

通过上面的配置，使用了 Struts2 的 FilterDispatcher 过滤所有的请求，这样调用的就是 Stuts2 的 MVC 框架，并且配置了 ContextLoaderListener。这样 Web 容器会自动加载 spring 的配置文件 applicationContext.xml，集成了 Spring 框架。

13.3.2　Spring 配置文件 applicationContext.xml 的配置

Spring 配置文件 applicationContext.xml 定义了所有要加载的类对象。通过此文件，实现了 Spring 与 Hibernate 的集成，并且所有的 Struts2 的 Action 类也要定义在这个文件中，去实现 Spring 和 Struts2 的集成，所以这个文件很重要。

```xml
<?xml version="1.0" encoding="utf-8"?>
<!-- 指定 Spring 配置文件的 Schema 信息 -->
<beans xmlns="http://www.springframework.org/schema/beans"
    xmlns:xsi="http://www.w3.org/2001/XMLSchema-instance"
    xmlns:aop="http://www.springframework.org/schema/aop"
    xmlns:tx="http://www.springframework.org/schema/tx"
    xsi:schemaLocation="http://www.springframework.org/schema/beans
    http://www.springframework.org/schema/beans/spring-beans-2.5.xsd
    http://www.springframework.org/schema/tx
    http://www.springframework.org/schema/tx/spring-tx-2.5.xsd
    http://www.springframework.org/schema/aop
    http://www.springframework.org/schema/aop/spring-aop-2.5.xsd">

    <!-- 定义数据源 Bean，使用 C3P0 数据源实现 -->
    <bean id="dataSource" class="com.mchange.v2.c3p0.ComboPooledDataSource"
        destroy-method="close">
        <!-- 指定连接数据库的驱动 -->
        <property name="driverClass" value="com.mysql.jdbc.Driver"/>
        <!-- 指定连接数据库的 URL -->
        <property name="jdbcUrl" value="jdbc:mysql://localhost:3306/bookstore"/>
        <!-- 指定连接数据库的用户名 -->
        <property name="user" value="root"/>
        <!-- 指定连接数据库的密码 -->
        <property name="password" value="1234"/>
        <!-- 指定连接数据库连接池的最大连接数 -->
        <property name="maxPoolSize" value="40"/>
        <!-- 指定连接数据库连接池的最小连接数 -->
        <property name="minPoolSize" value="1"/>
        <!-- 指定连接数据库连接池的初始化连接数 -->
        <property name="initialPoolSize" value="1"/>
        <!-- 指定连接数据库连接池的连接的最大空闲时间 -->
        <property name="maxIdleTime" value="20"/>
    </bean>

    <!-- 定义 Hibernate 的 SessionFactory -->
    <bean id="sessionFactory"
        class="org.springframework.crm.hibernate3.LocalSessionFactoryBean">
        <!-- 依赖注入数据源，注入正是上面定义的 dataSource -->
        <property name="dataSource" ref="dataSource"/>
```

```xml
        <!-- mappingResouces 属性用来列出全部映射文件 -->
        <property name="mappingResources">
            <list>
                <!-- 以下用来列出 Hibernate 映射文件 -->
                <value>iit/bookstore/entity/Bargain.hbm.xml</value>
                <value>iit/bookstore/entity/Book.hbm.xml</value>
                <value>iit/bookstore/entity/Manager.hbm.xml</value>
                <value>iit/bookstore/entity/Orders.hbm.xml</value>
                <value>iit/bookstore/entity/Recommended.hbm.xml</value>
                <value>iit/bookstore/entity/Sex.hbm.xml</value>
                <value>iit/bookstore/entity/Type.hbm.xml</value>
                <value>iit/bookstore/entity/User.hbm.xml</value>
                <value>iit/bookstore/entity/Ordersbook.hbm.xml</value>
            </list>
        </property>
        <!-- 定义 Hibernate 的 SessionFactory 的属性 -->
        <property name="hibernateProperties">
            <props>
                <!-- 指定数据库方言 -->
                <prop key="hibernate.dialect">
                    org.hibernate.dialect.MySQLInnoDBDialect</prop>
            </props>
        </property>
</bean>


<!--======================= 定义 DAO 组件 ========================= -->
<!-- 图书 Dao -->
<bean id="bookDao" class="iit.bookstore.dao.impl.BookDaoImpl">
    <property name="sessionFactory" ref="sessionFactory"/>
</bean>
<!-- 特价图书 Dao -->
<bean id="bargainDao" class="iit.bookstore.dao.impl.BargainDaoImpl">
    <property name="sessionFactory" ref="sessionFactory"/>
</bean>
<!-- 推荐图书 Dao -->
<bean id="recommendedDao" class="iit.bookstore.dao.impl.RecommendedDaoImpl">
    <property name="sessionFactory" ref="sessionFactory"/>
</bean>
<!-- 图书类别 Dao -->
<bean id="typeDao" class="iit.bookstore.dao.impl.TypeDaoImpl">
    <property name="sessionFactory" ref="sessionFactory"/>
</bean>
<!-- 管理员 Dao -->
<bean id="managerDao" class="iit.bookstore.dao.impl.ManagerDaoImpl">
    <property name="sessionFactory" ref="sessionFactory"/>
</bean>
<!-- 普通用户 Dao -->
<bean id="userDao" class="iit.bookstore.dao.impl.UserkDaoImpl">
```

```xml
        <property name="sessionFactory" ref="sessionFactory"/>
    </bean>
    <!-- 性别Dao -->
    <bean id="sexDao" class="iit.bookstore.dao.impl.SexDaoImpl">
        <property name="sessionFactory" ref="sessionFactory"/>
    </bean>
    <!-- 订单Dao -->
    <bean id="ordersDao" class="iit.bookstore.dao.impl.OrdersDaoImpl">
        <property name="sessionFactory" ref="sessionFactory"/>
    </bean>
    <!-- 订单图书Dao -->
    <bean id="ordersbookDao" class="iit.bookstore.dao.impl.OrdersbookDaoImpl">
        <property name="sessionFactory" ref="sessionFactory"/>
    </bean>


    <!--==================定义业务逻辑组件========================-->
    <!-- 图书管理 -->
    <bean id="bookManage" class="iit.bookstore.service.impl.BookManageImpl">
        <property name="bookDao" ref="bookDao"/>
        <property name="recommendedDao" ref="recommendedDao"/>
        <property name="bargainDao" ref="bargainDao"/>
        <property name="typeDao" ref="typeDao"/>
    </bean>
    <!-- 订单管理 -->
    <bean id="ordersManage" class="iit.bookstore.service.impl.OrdersManageImpl">
        <property name="ordersDao" ref="ordersDao"/>
        <property name="ordersbookDao" ref="ordersbookDao"/>
    </bean>
    <!-- 用户管理 -->
    <bean id="personManage" class="iit.bookstore.service.impl.PersonManageImpl">
        <property name="userDao" ref="userDao"/>
        <property name="managerDao" ref="managerDao"/>
        <property name="sexDao" ref="sexDao"/>
    </bean>

    <!--==================== 定义action组件=====================-->
    <!-- 用户注册action -->
    <bean id="EnrollAction" class="iit.bookstore.action.EnrollAction" scope="prototype">
        <property name="personManage" ref="personManage" />
    </bean>
    <!-- 用户登录action -->
    <bean id="LoginAction" class="iit.bookstore.action.LoginAction" scope="prototype">
        <property name="personManage" ref="personManage" />
    </bean>
    <!--网站首页action -->
    <bean name="MainAction" class="iit.bookstore.action.MainAction"
```

第13章 Spring+Struts2+Hibernate 集成实例

```xml
scope="prototype">
        <property name="bookManage" ref="bookManage" />
    </bean>

    <!-- 网站主页左侧边栏（销量排行榜）action -->
    <bean id="LeftAction" class="iit.bookstore.action.LeftAction" scope="prototype">
        <property name="bookManage" ref="bookManage" />
    </bean>

    <!--某一书籍的详细情况 -->
    <bean id="SingleBookAction" class="iit.bookstore.action.SingleBookAction" scope="prototype">
        <property name="bookManage" ref="bookManage" />
    </bean>

    <!--某一类栏目的所有书籍列表（最新上架、特价热销、精品推荐、销量排行榜） -->
    <bean id="OneTypeAction" class="iit.bookstore.action.OneTypeAction" scope="prototype">
        <property name="bookManage" ref="bookManage" />
    </bean>

    <!--购物车 -->
    <bean id="ShoppingCartAction" class="iit.bookstore.action.ShoppingCartAction" scope="prototype">
        <property name="bookManage" ref="bookManage" />
    </bean>

    <!--用户信息修改action -->
    <bean id="UserManageAction" class="iit.bookstore.action.UserManageAction" scope="prototype">
        <property name="personManage" ref="personManage" />
    </bean>
    <!--订单查看action -->
    <bean id="OrdersAction" class="iit.bookstore.action.OrdersAction" scope="prototype">
        <property name="ordersManage" ref="ordersManage" />
    </bean>
    <!--某一订单详情 -->
    <bean id="SingleOrdersAction" class="iit.bookstore.action.SingleOrdersAction" scope="prototype">
        <property name="ordersManage" ref="ordersManage" />
        <property name="bookManage" ref="bookManage" />
    </bean>
    <!--订单管理action -->
    <bean id="OrdersManageAction" class="iit.bookstore.action.OrdersManageAction" scope="prototype">
        <property name="ordersManage" ref="ordersManage" />
    </bean>
    <!-- 普通用户（管理员）登出 -->
```

```xml
        <bean id="LogoutAction" class="iit.bookstore.action.LogoutAction" scope="prototype">
        </bean>
        <!-- 管理员浏览图书列表 -->
        <bean name="ViewBooksAction" class="iit.bookstore.action.MainAction" scope="prototype">
            <property name="bookManage" ref="bookManage" />
        </bean>
        <!-- 管理员查看图书详细信息 -->
        <bean name="BookDetailAction" class="iit.bookstore.action.manage.BookDetailAction" scope="prototype">
            <property name="bookManage" ref="bookManage" />
        </bean>
        <!-- 图书增加 -->
        <bean id="BookAction" class="iit.bookstore.action.manage.BookAction" scope="prototype">
            <property name="bookManage" ref="bookManage" />
        </bean>
        <!-- 图书修改 -->
        <bean id="UpdateBookAction" class="iit.bookstore.action.manage.UpdateBookAction" scope="prototype">
            <property name="bookManage" ref ="bookManage" />
        </bean>
</beans>
```

13.3.3 Struts2 配置文件 Struts.xml 的配置

Struts2 配置文件 struts.xml 的配置如下：

```xml
<?xml version="1.0" encoding="GBK"?>
<!-- 指定 Struts2 配置文件的 DTD 信息 -->
<!DOCTYPE struts PUBLIC
    "-//Apache Software Foundation//DTD Struts Configuration 2.0//EN"
    "http://struts.apache.org/dtds/struts-2.0.dtd">
<!-- Struts2 配置文件的根元素 -->
<struts>
    <!-- 配置系列常量 -->
    <!-- 设置 Web 应用的默认编码集为 utf-8 -->
    <constant name="struts.i18n.encoding" value="utf-8" />
    <!-- 设置 Struts2 应用的国际化资源文件，多个文件中间可用逗号分隔 -->
    <constant name="struts.custom.i18n.resources" value="resource" />
    <!-- 设置 ognl 对静态方法的访问 -->
    <constant name="struts.ognl.allowStaticMethodAccess" value="true" />

    <package name="default" extends="struts-default">

        <!-- 配置自定义拦截器 LoginedCheckInterceptor -->
        <interceptors>
            <!-- 配置普通用户的权限检查拦截器 -->
            <interceptor name="userLoginCheck" class="iit.bookstore.action.interceptor.UserLoginCheck"></interceptor>
            <!-- 配置管理员的权限检查拦截器 -->
```

```xml
            <interceptor-stack name="userLoginStack">
                <interceptor-ref name="defaultStack" />
                <interceptor-ref name="userLoginCheck" />
            </interceptor-stack>
        </interceptors>

        <!-- 定义全局 Result 映射 -->
        <global-results>
            <!-- 定义login逻辑视图对应的视图资源 -->
            <result name="login">/login.jsp</result>
        </global-results>


        <!-- 用注册录 action -->
        <action name="enrollAction" class="EnrollAction">
            <result name="input">/enroll.jsp </result>
            <result name="success">/login.jsp</result>
        </action>
        <!-- 普通用户登录 action -->
        <action name="loginAction" class="LoginAction" method="loginCheck">
            <result name="error">/login.jsp </result>
            <result name="success" type="redirectAction">
                <param name="actionName">mainAction</param>
            </result>
        </action>
        <!-- 普通用户登出 -->
        <action name="logoutAction" class="LogoutAction">
            <result name="userLogout" type="redirect">/login.jsp </result>
        </action>


        <!--网站首页 action -->
        <action name="mainAction" class="MainAction">
            <result name="success">/main.jsp</result>
        </action>


        <!-- 网站主页左侧边栏（销量排行榜）action -->
        <action name="leftAction" class="LeftAction">
            <result>/leftResult.jsp</result>
        </action>


        <!--某一书籍的详细情况 -->
        <action name="singleBookAction" class="SingleBookAction">
            <result name="success">/singleBook.jsp</result>
        </action>
        <!--某一类栏目的所有书籍列表 -->
        <action name="oneTypeAction" class="OneTypeAction">
            <result name="error" type="redirect">/index.jsp </result>
            <result name="success">/oneType.jsp </result>
        </action>
        <!-- 购物车 action -->
        <action name="shoppingCartAction" class="ShoppingCartAction">
            <result name="success">/addToCart.jsp</result>
```

```xml
            <result name="login" type="redirect">/login.jsp </result>
            <interceptor-ref name="userLoginStack"></interceptor-ref>
        </action>
        <!-- 更改购物车中购物项 -->
        <action name="updateCartAction" class="ShoppingCartAction"
            method="updateCartItem">
            <result name="success">/shoppingCart.jsp</result>
            <result name="input">/shoppingCart.jsp </result>
            <interceptor-ref name="userLoginStack"></interceptor-ref>
        </action>
        <!-- 用户修改密码的 action -->
        <action name="updatePasswordAction" class="UserManageAction" method="updateUserPassword">
            <result name="success" type="redirect">/personalInformation.jsp </result>
            <result name="input">/updatePassword.jsp </result>
            <interceptor-ref name="userLoginStack"></interceptor-ref>
        </action>
        <!-- 用户修改信息的 action -->
        <action name="userInforAction" class="UserManageAction" method="updateUserInfor">
            <result name="success" type="redirect">/personalInformation.jsp </result>
            <result name="fail">/personalInformation.jsp </result>
            <interceptor-ref name="userLoginStack"></interceptor-ref>
        </action>
        <!-- 用户订单查看 action -->
        <action name="ordersAction" class="OrdersAction">
            <result name="success">/allOrders.jsp</result>
        </action>
        <!-- 某一订单的详情 -->
        <action name="singleOrdersAction" class="SingleOrdersAction">
            <result name="success">/singleOrders.jsp</result>
        </action>
        <!-- 用户订单更改 action -->
        <action name="ordersManageAction" class="OrdersManageAction">
            <result name="success" type="redirectAction">
                <param name="actionName">ordersAction</param>
            </result>
        </action>

    </package>

    <!--======================= 管理员的包======================= -->
    <package name="manage" extends="struts-default" namespace="/manage">
        <!-- 配置自定义拦截器 LoginedCheckInterceptor -->
        <interceptors>
            <!-- 配置管理员的权限检查拦截器 -->
            <interceptor name="managerLoginCheck" class="iit.bookstore.action.interceptor.ManagerLoginCheck"></interceptor>
            <interceptor-stack name="managerLoginStack">
                <interceptor-ref name="defaultStack" />
                <interceptor-ref name="managerLoginCheck" />
            </interceptor-stack>
```

第13章 Spring+Struts2+Hibernate 集成实例

```xml
        </interceptors>

        <!-- 定义全局 Result 映射 -->
        <global-results>
            <!-- 定义mannagerlogin逻辑视图对应的视图资源 -->
            <result name="managerLogin">/manage/index.jsp</result>
        </global-results>

        <!-- 管理员登录 action -->
        <action name="managerLoginAction" class="LoginAction" method="managerLoginCheck">
            <result name="error">/managerLogin.jsp </result>
            <result name="success">/manage/manageWelcome.jsp</result>
        </action>
        <!-- 管理员退出 -->
        <action name="logoutAction" class="LogoutAction">
            <result name="managerLogout" type="redirect">/managerLogin.jsp</result>
        </action>

        <!--浏览图书列表 -->
        <action name="viewBooksAction" class="ViewBooksAction">
            <result name="success">/manage/manageBook.jsp</result>
            <interceptor-ref name="managerLoginStack"/>
        </action>
        <!--浏览图书详细信息 -->
        <action name="bookDetailAction" class="BookDetailAction">
            <result name="success">/manage/singleBook.jsp</result>
            <interceptor-ref name="managerLoginStack"/>
        </action>

        <!-- 图书添加 -->
        <action name="bookAction" class="BookAction">
            <!-- 动态设置Action的属性值，设置文件上传后的保存位置 -->
            <param name="savePath">/upload</param>
            <result name="input">/manage/addBook.jsp </result>
            <result name="success" type="redirect">/manage/addBook.jsp</result>
            <!-- 配置fileUpload的拦截器 -->
            <interceptor-ref name="fileUpload">
                <!-- 允许上传的文件类型 -->
                <param name="allowedTypes">
                    image/bmp,image/png,image/gif,image/jpeg,image/jpg,image/x-png,image/pjpeg
                </param>
                <!-- 允许上传的文件大小 -->
                <param name="maximumSize">2000000</param>
            </interceptor-ref>

            <interceptor-ref name="managerLoginStack"/>
        </action>

        <!-- 图书修改 -->
```

```xml
<action name="updateBookAction" class="UpdateBookAction">
    <!-- 动态设置Action的属性值，设置文件上传后的保存位置 -->
    <param name="savePath">/upload</param>
    <result name="input">/manage/updateBook.jsp</result>
    <result name="success">/manage/singleBook.jsp</result>
    <result name="preUpdate">/manage/updateBook.jsp</result>

    <interceptor-ref name="fileUpload">
        <param name="allowedTypes">
            image/bmp,image/png,image/gif,image/jpeg,image/jpg,image/x-png,image/pjpeg </param>
        <param name="maximumSize">2000000</param>
    </interceptor-ref>
    <interceptor-ref name="managerLoginStack"/>
</action>
    </package>
</struts>
```

13.3.4 国际化资源文件的配置

Struts2 国际化资源文件 globalMessages_zh_CN.properties 的配置代码如下：

```
struts.messages.error.content.type.not.allowed=\u4E0A\u4F20\u6587\u4EF6\u5FC5\u987B\u4E3A\u56FE\u7247
struts.messages.error.file.too.large=\u4E0A\u4F20\u6587\u4EF6\u592A\u5927
struts.messages.error.uploading=\u4E0A\u4F20\u8FC7\u7A0B\u51FA\u73B0\u5F02\u5E38\uFF0C\u8BF7\u91CD\u8BD5
```

13.4 编写持久化对象（PO）

面向对象的分析，是指根据系统的需求提取应用中的对象，将这些对象抽象成类，这些需要保存的持久化类就是持久化对象（PO），即JavaBean组件。数据库中有9张表，也即有9个JavaBean组件需要持久化，这些持久化类分别为：Book类、Type类、Bargain类、Recommended类、Orders类、Ordersbook类、Manager类、User类和Sex类，要将这些持久化类交给Hibernate管理，必须定义相应的映射文件。持久化类及其映射文件在系统中的结构图如图13-2所示：

这里，只把最重要的持久化类及其映射文件列举出来，其他的类及映射文件信息的编写方法与此相似，由于篇幅所限，这里只列举出重要的几个持久化类及映射文件。

图13-2 系统的Javabean组件及映射文件的结构图

第 13 章　Spring+Struts2+Hibernate 集成实例

13.4.1　定义 Book 类及映射文件

Book 类的代码如下：

```java
package iit.bookstore.entity;
import java.sql.*;

@SuppressWarnings("serial")
public class Book implements java.io.Serializable {
    private Integer bookId;
    private String bookNumber;
    private String bookName;
    private String bookAuthor;
    private String bookPress;
    private String bookPicture;
    private Integer bookAmount;
    private Type type;
    private Timestamp bookShelveTime;
    private Double bookPrice;
    private String bookRemark;
    private Integer bookSales;
    private Double bookNewPrice;

    public Book() {
    }

    public Book(String bookNumber, String bookName, String bookAuthor,
            String bookPress, String bookPicture, Integer bookAmount,
            Type type, Timestamp bookShelveTime, Double bookPrice,
            Integer bookSales) {
        this.bookNumber = bookNumber;
        this.bookName = bookName;
        this.bookAuthor = bookAuthor;
        this.bookPress = bookPress;
        this.bookPicture = bookPicture;
        this.bookAmount = bookAmount;
        this.type = type;
        this.bookShelveTime = bookShelveTime;
        this.bookPrice = bookPrice;
        this.bookSales = bookSales;
    }

    public Book(String bookNumber, String bookName, String bookAuthor,
            String bookPress, String bookPicture, Integer bookAmount,
            Type type, Timestamp bookShelveTime, Double bookPrice,
            String bookRemark, Integer bookSales,Double bookNewPrice) {
        this.bookNumber = bookNumber;
        this.bookName = bookName;
        this.bookAuthor = bookAuthor;
        this.bookPress = bookPress;
```

```java
        this.bookPicture = bookPicture;
        this.bookAmount = bookAmount;
        this.type = type;
        this.bookShelveTime = bookShelveTime;
        this.bookPrice = bookPrice;
        this.bookRemark = bookRemark;
        this.bookSales = bookSales;
        this.bookNewPrice = bookNewPrice;
    }

    public Integer getBookId() {
        return this.bookId;
    }

    public void setBookId(Integer bookId) {
        this.bookId = bookId;
    }

    public String getBookNumber() {
        return this.bookNumber;
    }

    public void setBookNumber(String bookNumber) {
        this.bookNumber = bookNumber;
    }

    public String getBookName() {
        return this.bookName;
    }

    public void setBookName(String bookName) {
        this.bookName = bookName;
    }

    public String getBookAuthor() {
        return this.bookAuthor;
    }

    public void setBookAuthor(String bookAuthor) {
        this.bookAuthor = bookAuthor;
    }

    public String getBookPress() {
        return this.bookPress;
    }

    public void setBookPress(String bookPress) {
        this.bookPress = bookPress;
    }
```

第 13 章　Spring+Struts2+Hibernate 集成实例

```java
public String getBookPicture() {
    return this.bookPicture;
}

public void setBookPicture(String bookPicture) {
    this.bookPicture = bookPicture;
}

public Integer getBookAmount() {
    return this.bookAmount;
}

public void setBookAmount(Integer bookAmount) {
    this.bookAmount = bookAmount;
}

public Type getType() {
    return this.type;
}

public void setType(Type type) {
    this.type = type;
}

public Timestamp getBookShelveTime() {
    return this.bookShelveTime;
}

public void setBookShelveTime(Timestamp bookShelveTime) {
    this.bookShelveTime = bookShelveTime;
}

public Double getBookPrice() {
    return this.bookPrice;
}

public void setBookPrice(Double bookPrice) {
    this.bookPrice = bookPrice;
}

public String getBookRemark() {
    return this.bookRemark;
}

public void setBookRemark(String bookRemark) {
    this.bookRemark = bookRemark;
}

public Integer getBookSales() {
    return this.bookSales;
```

```java
    }

    public void setBookSales(Integer bookSales) {
        this.bookSales = bookSales;
    }

    public Double getBookNewPrice() {
        return bookNewPrice;
    }

    public void setBookNewPrice(Double bookNewPrice) {
        this.bookNewPrice = bookNewPrice;
    }

    @Override
    public int hashCode() {
        final int prime = 31;
        int result = 1;
        result = prime * result + ((bookId == null) ? 0 : bookId.hashCode());
        return result;
    }

    @Override
    public boolean equals(Object obj) {
        if(this == obj)
            return true;
        if(obj == null)
            return false;
        if(getClass() != obj.getClass())
            return false;
        Book other = (Book) obj;
        if(bookId == null) {
            if(other.bookId != null)
                return false;
        } else if(!bookId.equals(other.bookId))
            return false;
        return true;
    }
}
```

Book 类映射文件 Book.hbm.xml 的代码如下：

```xml
<?xml version="1.0" encoding="utf-8"?>
<!DOCTYPE hibernate-mapping PUBLIC "-//Hibernate/Hibernate Mapping DTD 3.0//EN"
    "http://hibernate.sourceforge.net/hibernate-mapping-3.0.dtd">

<hibernate-mapping package="iit.bookstore.entity">
 <class catalog="bookstore" name="Book" table="book">
  <id name="bookId" type="java.lang.Integer">
```

```xml
    <column name="bookId"/>
    <generator class="native"/>
  </id>
  <many-to-one class="Type" lazy="false" name="type">
    <column name="typeId" not-null="true"/>
  </many-to-one>
  <property generated="never" lazy="false" name="bookNumber" type="java.lang.String">
    <column length="21" name="bookNumber" not-null="true"/>
  </property>
  <property generated="never" lazy="false" name="bookName" type="java.lang.String">
    <column length="20" name="bookName" not-null="true"/>
  </property>
  <property generated="never" lazy="false" name="bookAuthor" type="java.lang.String">
    <column length="20" name="bookAuthor" not-null="true"/>
  </property>
  <property generated="never" lazy="false" name="bookPress" type="java.lang.String">
    <column length="20" name="bookPress" not-null="true"/>
  </property>
  <property generated="never" lazy="false" name="bookPicture" type="java.lang.String">
    <column length="100" name="bookPicture" not-null="true"/>
  </property>
  <property generated="never" lazy="false" name="bookAmount" type="java.lang.Integer">
    <column name="bookAmount" not-null="true"/>
  </property>
  <property generated="never" lazy="false" name="bookShelveTime" type="timestamp">
    <column length="14" name="bookShelveTime" not-null="true"/>
  </property>
  <property generated="never" lazy="false" name="bookPrice" type="java.lang.Double">
    <column name="bookPrice" not-null="true" precision="10"/>
  </property>
  <property generated="never" lazy="false" name="bookRemark" type="java.lang.String">
    <column length="200" name="bookRemark"/>
  </property>
  <property generated="never" lazy="false" name="bookSales" type="java.lang.Integer">
    <column name="bookSales" not-null="true"/>
  </property>
  </class>
</hibernate-mapping>
```

13.4.2 定义 Bargain 类及映射文件

Bargain 类的代码如下：

```java
package iit.bookstore.entity;

@SuppressWarnings("serial")
public class Bargain implements java.io.Serializable {

    private Integer bargainId;
    private Book book;
    private Double bookNewPrice;

    public Bargain() {
    }

    public Bargain(Book book, Double bookNewPrice) {
        this.book = book;
        this.bookNewPrice = bookNewPrice;
    }

    public Integer getBargainId() {
        return this.bargainId;
    }

    public void setBargainId(Integer bargainId) {
        this.bargainId = bargainId;
    }

    public Book getBook() {
        return book;
    }

    public void setBook(Book book) {
        this.book = book;
    }

    public Double getBookNewPrice() {
        return this.bookNewPrice;
    }

    public void setBookNewPrice(Double bookNewPrice) {
        this.bookNewPrice = bookNewPrice;
    }
}
```

Bargain 类映射文件 Bargain.hbm.xml 的代码如下：

```xml
<?xml version="1.0" encoding="utf-8"?>
<!DOCTYPE hibernate-mapping PUBLIC "-//Hibernate/Hibernate Mapping DTD 3.0//EN"
```

```xml
"http://hibernate.sourceforge.net/hibernate-mapping-3.0.dtd">

<hibernate-mapping package="iit.bookstore.entity">
    <class name="Bargain" table="bargain" catalog="bookstore">
        <id name="bargainId" type="java.lang.Integer">
            <column name="bargainId" />
            <generator class="native" />
        </id>
        <many-to-one name="book" class="Book" lazy="false" unique="true">
            <column name="bookId" not-null="true" />
        </many-to-one>
        <property name="bookNewPrice" type="java.lang.Double">
            <column name="bookNewPrice" precision="10" not-null="true" />
        </property>
    </class>
</hibernate-mapping>
```

13.4.3 定义 Orders 类及映射文件

Orders 类的代码如下：

```java
package iit.bookstore.entity;
import java.sql.Timestamp;
@SuppressWarnings("serial")
public class Orders implements java.io.Serializable {

    /** 订单编号 */
    private Integer ordersId;
    /** 订单号 */
    private String ordersNumber;
    /** 用户 */
    private User user;
    /** 下单时间 */
    private Timestamp ordersTime;
    /** 订单号完成状态 */
    private String isDeal;
    /** 订单金额 */
    private Double totalMoney;

    public Orders() {
    }

    public Orders(String ordersNumber,User user, Timestamp ordersTime,String isDeal,Double totalMoney) {
        this.ordersNumber = ordersNumber;
        this.user = user;
        this.ordersTime = ordersTime;
        this.isDeal = isDeal;
        this.totalMoney = totalMoney;
    }
```

```java
    public Integer getOrdersId() {
        return this.ordersId;
    }

    public void setOrdersId(Integer ordersId) {
        this.ordersId = ordersId;
    }

    public String getOrdersNumber() {
        return ordersNumber;
    }

    public void setOrdersNumber(String ordersNumber) {
        this.ordersNumber = ordersNumber;
    }

    public User getUser() {
        return user;
    }

    public void setUser(User user) {
        this.user = user;
    }

    public Timestamp getOrdersTime() {
        return this.ordersTime;
    }

    public void setOrdersTime(Timestamp ordersTime) {
        this.ordersTime = ordersTime;
    }

    public String getIsDeal() {
        return isDeal;
    }

    public void setIsDeal(String isDeal) {
        this.isDeal = isDeal;
    }

    public Double getTotalMoney() {
        return totalMoney;
    }

    public void setTotalMoney(Double totalMoney) {
        this.totalMoney = totalMoney;
    }

}
```

Orders 类映射文件 Orders.hbm.xml 的代码如下:

```xml
<?xml version="1.0" encoding="utf-8"?>
<!DOCTYPE hibernate-mapping PUBLIC "-//Hibernate/Hibernate Mapping DTD 3.0//EN"
"http://hibernate.sourceforge.net/hibernate-mapping-3.0.dtd">

<hibernate-mapping package="iit.bookstore.entity">
    <class name="Orders" table="orders" catalog="bookstore">
        <id name="ordersId" type="java.lang.Integer">
            <column name="ordersId" />
            <generator class="native" />
        </id>
        <property name="ordersNumber" type="java.lang.String">
            <column name="ordersNumber" length="21" not-null="true" />
        </property>
        <many-to-one name="user" class="User" lazy="false">
            <column name="userId" not-null="true" />
        </many-to-one>
        <property name="ordersTime" type="timestamp">
            <column name="ordersTime" length="19" not-null="true" />
        </property>
        <property name="isDeal" type="java.lang.String">
            <column name="isDeal" length="1" not-null="true" />
        </property>
        <property name="totalMoney" type="java.lang.Double">
            <column name="totalMoney" precision="10" not-null="true" />
        </property>
    </class>
</hibernate-mapping>
```

13.4.4 定义 Ordersbook 类及映射文件

Ordersbook 类的代码如下:

```java
package iit.bookstore.entity;
@SuppressWarnings("serial")
public class Ordersbook implements java.io.Serializable {
    /** ID */
    private Integer ordersBookId;
    /** 订单 */
    private Orders orders;
    /** 图书 */
    private Book book;
    /** 图书数量 */
    private Integer bookAmount;

    public Ordersbook() {
    }

    public Ordersbook(Orders orders, Book book, Integer bookAmount) {
```

```java
        this.orders = orders;
        this.book = book;
        this.bookAmount = bookAmount;
    }

    public Integer getOrdersBookId() {
        return this.ordersBookId;
    }

    public void setOrdersBookId(Integer ordersBookId) {
        this.ordersBookId = ordersBookId;
    }

    public Orders getOrders() {
        return orders;
    }

    public void setOrders(Orders orders) {
        this.orders = orders;
    }

    public Book getBook() {
        return book;
    }

    public void setBook(Book book) {
        this.book = book;
    }

    public Integer getBookAmount() {
        return this.bookAmount;
    }

    public void setBookAmount(Integer bookAmount) {
        this.bookAmount = bookAmount;
    }

}
```

Ordersbook 类映射文件 Ordersbook.hbm.xml 的代码如下：

```xml
<?xml version="1.0" encoding="utf-8"?>
<!DOCTYPE hibernate-mapping PUBLIC "-//Hibernate/Hibernate Mapping DTD 3.0//EN"
"http://hibernate.sourceforge.net/hibernate-mapping-3.0.dtd">

<hibernate-mapping package="iit.bookstore.entity">
    <class name="Ordersbook" table="ordersbook" catalog="bookstore">
        <id name="ordersBookId" type="java.lang.Integer">
            <column name="ordersBookId" />
            <generator class="native" />
```

```xml
        </id>
        <many-to-one name="orders" class="Orders" lazy="false">
            <column name="ordersId" not-null="true" />
        </many-to-one>
        <many-to-one name="book" class="Book" lazy="false" unique="true">
            <column name="bookId" not-null="true" />
        </many-to-one>
        <property name="bookAmount" type="java.lang.Integer">
            <column name="bookAmount" not-null="true" />
        </property>
    </class>
</hibernate-mapping>
```

13.4.5 定义 User 类及映射文件

User 类的代码如下:

```java
package iit.bookstore.entity;

@SuppressWarnings("serial")
public class User implements java.io.Serializable {

    private Integer userId;
    private String userName;
    private String userPassword;
    private String userEmail;
    private String userNickname;
    private Sex sex;
    private String userAddress;
    private String userPhone;
    private String userRemark;

    public User() {
    }

    public User(String userName, String userPassword, String userEmail,
            Sex sex) {
        this.userName = userName;
        this.userPassword = userPassword;
        this.userEmail = userEmail;
        this.sex = sex;
    }

    public User(String userName, String userPassword, String userEmail,
            String userNickname, Sex sex, String userAddress,
            String userPhone, String userRemark) {
        this.userName = userName;
        this.userPassword = userPassword;
        this.userEmail = userEmail;
        this.userNickname = userNickname;
```

```java
        this.sex = sex;
        this.userAddress = userAddress;
        this.userPhone = userPhone;
        this.userRemark = userRemark;
    }

    public Integer getUserId() {
        return this.userId;
    }

    public void setUserId(Integer userId) {
        this.userId = userId;
    }

    public String getUserName() {
        return this.userName;
    }

    public void setUserName(String userName) {
        this.userName = userName;
    }

    public String getUserPassword() {
        return this.userPassword;
    }

    public void setUserPassword(String userPassword) {
        this.userPassword = userPassword;
    }

    public String getUserEmail() {
        return this.userEmail;
    }

    public void setUserEmail(String userEmail) {
        this.userEmail = userEmail;
    }

    public String getUserNickname() {
        return this.userNickname;
    }

    public void setUserNickname(String userNickname) {
        this.userNickname = userNickname;
    }

    public Sex getSex() {
        return sex;
    }
```

```java
    public void setSex(Sex sex) {
        this.sex = sex;
    }

    public String getUserAddress() {
        return this.userAddress;
    }

    public void setUserAddress(String userAddress) {
        this.userAddress = userAddress;
    }

    public String getUserPhone() {
        return this.userPhone;
    }

    public void setUserPhone(String userPhone) {
        this.userPhone = userPhone;
    }

    public String getUserRemark() {
        return this.userRemark;
    }

    public void setUserRemark(String userRemark) {
        this.userRemark = userRemark;
    }
}
```

User 类映射文件 User.hbm.xml 的代码如下：

```xml
<?xml version="1.0" encoding="utf-8"?>
<!DOCTYPE hibernate-mapping PUBLIC "-//Hibernate/Hibernate Mapping DTD 3.0//EN"
"http://hibernate.sourceforge.net/hibernate-mapping-3.0.dtd">

<hibernate-mapping package="iit.bookstore.entity">
    <class name="User" table="user" catalog="bookstore">
        <id name="userId" type="java.lang.Integer">
            <column name="userId" />
            <generator class="native" />
        </id>
        <many-to-one name="sex" class="Sex" lazy="false">
            <column name="sexId" not-null="true" />
        </many-to-one>
        <property name="userName" type="java.lang.String">
            <column name="userName" length="16" not-null="true" />
        </property>
        <property name="userPassword" type="java.lang.String">
            <column name="userPassword" length="12" not-null="true" />
```

```xml
        </property>
        <property name="userEmail" type="java.lang.String">
            <column name="userEmail" length="100" not-null="true" />
        </property>
        <property name="userNickname" type="java.lang.String">
            <column name="userNickname" length="10" />
        </property>
        <property name="userAddress" type="java.lang.String">
            <column name="userAddress" length="200" />
        </property>
        <property name="userPhone" type="java.lang.String">
            <column name="userPhone" length="24" />
        </property>
        <property name="userRemark" type="java.lang.String">
            <column name="userRemark" length="200" />
        </property>
    </class>
</hibernate-mapping>
```

13.5 建立数据库访问层组件（DAO）

在 Hibernate 持久层之上，可以使用 DAO 组件再次封装数据库操作，这是 Java EE 应用里常用的 DAO 模式。DAO 层之上是业务逻辑层，业务逻辑层组件负责业务逻辑的变化，而 DAO 层组件负责持久化技术的变化。每个 DAO 组件都包含了对数据库的逻辑访问，可对一个数据表完成基本的增、删、查、改操作。对于不同的持久化技术，Spring 的 DAO 提供了一个 DAO 模板，将通用的操作都放在模板里完成，这样就减少了代码编写量和代码编写所带来的人为错误。

13.5.1 DAO 组件接口的定义

DAO 组件提供了各种持久化对象基本的增、删、查、改操作（即 CRUD 操作）。为了避免业务逻辑组件和特定的 DAO 组件耦合，在开发应用程序时都采用面向接口的编程：在 DAO 的接口中对提供的各种 CRUD 操作提供声明，而无须给出任何的实现。DAO 组件的具体实现由其实现类来完成，实现类中编写 CRUD 方法的具体代码。该网上书店系统中共有 9 个持久化了，所以其 DAO 组件也有 9 个。相应地这里只给出重要的 DAO 组建的声明，其他的 DAO 组件省略。下面是关于这些 DAO 组件接口的源代码。

BookDao 接口的定义如下：

```java
package iit.bookstore.dao;

import iit.bookstore.entity.Book;

import java.util.List;

public interface BookDao {

    /** 添加新的图书 */
```

```java
    public int add(Book book);
    
    /** 修改图书 */
    public void update(Book book);
    
    /** 根据图书ID查询该图书 */
    public Book findById(int bookId);
    
    /** 查询销量最好的图书 */
    public List<Book> bestSellingBook(int pageNumber, int pageSize);
    
    /** 查询最新上架的图书 */
    public List<Book> latestBook(int pageNumber, int pageSize);
    
    /** 根据图书名称查询图书 */
    public List<Book> allBookByName(String bookName, int pageNumber,
            int pageSize);
    
    /** 根据作者查询图书 */
    public List<Book> allBookByAuthor(String bookAuthor, int pageNumber,
            int pageSize);
    
    /** 根据出版社查询图书 */
    public List<Book> allBookByPress(String bookPress, int pageNumber,
            int pageSize);
    
    /** 根据图书类别来查询图书 */
    public List<Book> allBookByType(int typeId, int pageNumber, int pageSize);
}
```

BargainDao 接口的定义如下：

```java
package iit.bookstore.dao;

import iit.bookstore.entity.Bargain;

import java.util.List;

public interface BargainDao {
    /** 查询特价图书ID */
    public List<Bargain> findByPage(int pageNumber, int pageSize);
    /** 查询图书是否为特价图书,若不是则返回null*/
    public Bargain isBargain(int bookId);
}
```

OrdersDao 接口的定义如下：

```java
package iit.bookstore.dao;

import iit.bookstore.entity.Orders;
```

```java
import java.util.List;

public interface OrdersDao {

    /** 添加一个新的订单 */
    public int add(Orders orders);

    /** 删除一个订单 */
    public void delete(int ordersId);

    /** 修改订单 */
    public void update(Orders orders);

    /** 根据ID查询订单 */
    public Orders findById(int ordersId);

    /** 查询所有订单 */
    public List<Orders> findAll(final int pageNumber, final int pageSize);

    /** 查询所有订单-依据状态 */
    public List<Orders> allOrdersByDeal(final String isDeal,
            final int pageNumber, final int pageSize);

    /** 根据userId获取该用户所有订单 */
    public List<Orders> allOrdersByUser(int userId, int pageNumber, int pageSize);

    /** 根据userId,订单处理状态获取该用户所有订单 */
    public List<Orders> allOrdersByUserDeal(int userId, String isDeal,
            int pageNumber, int pageSize);

}
```

OrdersbookDao 接口的定义如下:

```java
package iit.bookstore.dao;

import iit.bookstore.entity.Ordersbook;

import java.util.List;

public interface OrdersbookDao {

    /** 添加一条订单图书信息 */
    public void add(Ordersbook ordersbook);

    /**根据ordersId获取该订单所有订单图书 */
    public List<Ordersbook> allOrdersbookByOrders(int ordersId);

}
```

UserDao 接口的定义如下:

第 13 章　Spring+Struts2+Hibernate 集成实例

```java
package iit.bookstore.dao;

import iit.bookstore.entity.User;

public interface UserDao {

    /** 获取用户信息 */
    public User findById(int userId);

    /** 通过username和password查询User对象，若有则返回User实例，否则null*/
    public User findByNamePasswd(String userName, String userPassword);

    /** 通过username查询User对象，若有则返回User实例，否则null*/
    public User findByName(String userName);

    /** 添加一条用户信息 */
    public int add(User user);

    /** 修改一条用户信息 */
    public void update(User user);
}
```

13.5.2　实现 DAO 组件

借助 Spring 的 DAO 支持，可以很方便地为 DAO 接口提供实现类，Spring 为 Hibernate 提供的 DAO 基类是 HibernateDaoSupport，该类只须传入一个 SessionFactory 应用，即可得到一个 HibernateTemplate 实例，该实例可以很容易地实现数据的大部分操作。

本网上书店系统还扩展了 HibernateDaoSupport，提供了一个 ExtendHibernateDaoSupport 子类，该子类完成了大量的分页查询的方法，可以更好地完成分页查询。该类具体实现代码如下：

```java
package iit.bookstore.dao.base;

import org.springframework.orm.hibernate3.HibernateCallback;
import org.springframework.orm.hibernate3.support.HibernateDaoSupport;
import org.hibernate.Session;
import org.hibernate.Query;
import org.hibernate.HibernateException;
import java.sql.SQLException;
import java.util.List;

/**
 * 对Hibernate框架方法的扩展
 * @author Administrator
 *
 */
public class ExtendHibernateDaoSupport extends HibernateDaoSupport
{
```

```java
/**
 * 使用 HQL 语句进行分页查询
 * @param hql 需要查询的 HQL 语句
 * @param offset 第一条记录索引
 * @param pageSize 每页需要显示的记录数
 * @return 当前页的所有记录
 */
@SuppressWarnings("rawtypes")
public List findByPage(final String hql,
    final int offset, final int pageSize)
{
    //通过一个 HibernateCallback 对象来执行查询
    List list = getHibernateTemplate()
        .executeFind(new HibernateCallback()
    {
        //实现 HibernateCallback 接口必须实现的方法
        public Object doInHibernate(Session session)
            throws HibernateException, SQLException
        {
            //执行 Hibernate 分页查询
            List result = session.createQuery(hql)
                .setFirstResult(offset)
                .setMaxResults(pageSize)
                .list();
            return result;
        }
    });
    return list;
}

/**
 * 使用 HQL 语句进行分页查询
 * @param hql 需要查询的 HQL 语句
 * @param value 如果 HQL 有一个参数需要传入，value 就是传入 hql 语句的参数
 * @param offset 第一条记录索引
 * @param pageSize 每页需要显示的记录数
 * @return 当前页的所有记录
 */
@SuppressWarnings("rawtypes")
public List findByPage(final String hql , final Object value ,
    final int offset, final int pageSize)
{
    //通过一个 HibernateCallback 对象来执行查询
    List list = getHibernateTemplate()
        .executeFind(new HibernateCallback()
    {
        //实现 HibernateCallback 接口必须实现的方法
        public Object doInHibernate(Session session)
            throws HibernateException, SQLException
        {
```

```java
            //执行Hibernate分页查询
            List result = session.createQuery(hql)
                //为HQL语句传入参数
                .setParameter(0, value)
                .setFirstResult(offset)
                .setMaxResults(pageSize)
                .list();
            return result;
        }
    });
    return list;
}

/**
 * 使用HQL语句进行分页查询
 * @param HQL 需要查询的HQL语句
 * @param values 如果HQL有多个个参数需要传入，values就是传入hql的参数数组
 * @param offset 第一条记录索引
 * @param pageSize 每页需要显示的记录数
 * @return 当前页的所有记录
 */
@SuppressWarnings("rawtypes")
public List findByPage(final String hql, final Object[] values,
    final int offset, final int pageSize)
{
    //通过一个HibernateCallback对象来执行查询
    List list = getHibernateTemplate()
        .executeFind(new HibernateCallback()
        {
            //实现HibernateCallback接口必须实现的方法
            public Object doInHibernate(Session session)
                throws HibernateException, SQLException
            {
                //执行Hibernate分页查询
                Query query = session.createQuery(hql);
                //为HQL语句传入参数
                for(int i = 0 ; i < values.length ; i++)
                {
                    query.setParameter( i, values[i]);
                }
                List result = query.setFirstResult(offset)
                    .setMaxResults(pageSize)
                    .list();
                return result;
            }
        });
    return list;
}
}
```

应用中实际的 DAO 实现类都要继承 HibernateDaoSupport，并实现相应的接口，而我们的系统中的 DAO 实现类只要继承子类，可以更轻松地完成包括分页查询在的各种方法。

如下是 BookDao 组件的具体实现类 BookDaoImpl 的源代码：

```java
package iit.bookstore.dao.impl;

import java.util.List;

import iit.bookstore.dao.BookDao;
import iit.bookstore.dao.base.ExtendHibernateDaoSupport;
import iit.bookstore.entity.Book;

public class BookDaoImpl extends ExtendHibernateDaoSupport implements BookDao{

    /** 添加新的图书 */
    public int add(Book book) {
        return (Integer)getHibernateTemplate().save(book);
    }
    /** 修改图书 */
    public void update(Book book) {
        getHibernateTemplate().update(book);
    }

    /** 根据图书 ID 查询该图书 */
    public Book findById(int bookId) {
        return (Book)getHibernateTemplate().get(Book.class,bookId);
    }

    /** 查询销量最好的图书 */
    @SuppressWarnings("unchecked")
    public List<Book> bestSellingBook(int pageNumber, int pageSize) {
        return (List<Book>)getHibernateTemplate().find(
                "from Book as book where book.bookSales > 0 and book.bookAmount > 0 order by book.bookSales desc");

    }

    /** 查询最新上架的图书 */
    @SuppressWarnings("unchecked")
    public List<Book> latestBook(int pageNumber, int pageSize) {
        return (List<Book>) findByPage("from Book as book where book.bookAmount > 0 order by book.bookShelveTime desc ", pageNumber*pageSize, pageSize);
    }

    /** 根据图书名称查询图书 */
    @SuppressWarnings("unchecked")
    public List<Book> allBookByName(String bookName, int pageNumber,int
```

```java
pageSize) {
        return findByPage("from Book as book where book.bookName like
'%" + bookName + "%'",pageNumber*pageSize , pageSize);
    }

    /** 根据作者查询图书 */
    @SuppressWarnings("unchecked")
    public List<Book> allBookByAuthor(String bookAuthor, int pageNumber,
int pageSize) {
        return findByPage("from Book as book where book.bookAuthor like '%"
+ bookAuthor + "%'",pageNumber*pageSize , pageSize);
    }

    /** 根据出版社查询图书 */
    @SuppressWarnings("unchecked")
    public List<Book> allBookByPress(String bookPress, int pageNumber, int
pageSize) {
        return findByPage("from Book as book where book.bookPress like '%"
+ bookPress + "%'",pageNumber*pageSize,pageSize);
    }

    /** 根据图书类别来查询图书 */
    @SuppressWarnings("unchecked")
    public List<Book> allBookByType(int typeId, int pageNumber, int
pageSize){
        return findByPage("from Book as book where
book.type.typeId=?",typeId,pageNumber*pageSize,pageSize);
    }
}
```

如下是 BargainDao 组件的具体实现类 BargainDaoImpl 的源代码：

```java
package iit.bookstore.dao.impl;

import java.util.List;

import iit.bookstore.dao.BargainDao;
import iit.bookstore.dao.base.ExtendHibernateDaoSupport;
import iit.bookstore.entity.Bargain;

public class BargainDaoImpl extends ExtendHibernateDaoSupport implements
BargainDao{

    /** 查询特价图书 ID */
    @SuppressWarnings("unchecked")
    public List<Bargain> findByPage(int pageNumber, int pageSize) {
        return findByPage("from Bargain",pageNumber*pageSize,pageSize);
    }
```

```java
        /** 查询图书是否为特价图书,若不是则返回 null*/
        @SuppressWarnings("unchecked")
        public Bargain isBargain(int bookId){
                List<Bargain> bargainList = getHibernateTemplate().find("from Bargain as bargain where bargain.book.bookId= ?",bookId);
                return bargainList.size() > 0 ? bargainList.get(0) : null;
        }
}
```

如下是 OrdersDao 组件的具体实现类 OrdersDaoImpl 的源代码:

```java
package iit.bookstore.dao.impl;

import java.util.List;

import iit.bookstore.dao.OrdersDao;
import iit.bookstore.dao.base.ExtendHibernateDaoSupport;
import iit.bookstore.entity.Orders;

public class OrdersDaoImpl extends ExtendHibernateDaoSupport implements OrdersDao{
    /** 添加一个新的订单 */
    public int add(Orders orders){
        return (Integer)getHibernateTemplate().save(orders);
    }

    /** 删除一个订单 */
    public void delete(int ordersId){
        getHibernateTemplate().delete(findById(ordersId));
    }

    /** 修改订单 */
    public void update(Orders orders){
        getHibernateTemplate().update(orders);
    }

    /** 根据 ID 查询订单 */
    public Orders findById(int ordersId){
        return (Orders)getHibernateTemplate().get(Orders.class, ordersId);
    }
    /** 查询所有订单 */
    @SuppressWarnings("unchecked")
    public List<Orders> findAll(final int pageNumber,final int pageSize){
        return findByPage("from Orders as orders order by ordersTime desc", pageNumber*pageSize, pageSize);
    }

    /** 查询所有订单-依据状态 */
    @SuppressWarnings("unchecked")
    public List<Orders> allOrdersByDeal(final String isDeal,final int pageNumber,final int pageSize){
```

第 13 章　Spring+Struts2+Hibernate 集成实例

```java
        return findByPage("from Orders as orders where orders.isDeal = ? order by ordersTime desc", isDeal, pageNumber*pageSize, pageSize);
    }

    /** 根据 userId 获取该用户所有订单 */
    @SuppressWarnings("unchecked")
    public List<Orders> allOrdersByUser(int userId,int pageNumber,int pageSize){
        return findByPage("from Orders as orders where orders.user.userId= ? order by ordersTime desc",
            userId,pageNumber*pageSize,pageSize);
    }
    /** 根据 userId,订单处理状态获取该用户所有订单 */
    @SuppressWarnings("unchecked")
    public List<Orders> allOrdersByUserDeal(int userId,String isDeal,int pageNumber,int pageSize){
        return findByPage("from Orders as orders where orders.user.userId= ? and orders.isDeal= ? order by ordersTime desc",
            new Object[]{userId,isDeal},pageNumber*pageSize,pageSize);
    }
}
```

如下是 OrdersbookDao 组件的具体实现类 OrdersbookDaoImpl 的源代码：

```java
package iit.bookstore.dao.impl;

import java.util.List;

import iit.bookstore.dao.OrdersbookDao;
import iit.bookstore.dao.base.ExtendHibernateDaoSupport;
import iit.bookstore.entity.Ordersbook;

public class OrdersbookDaoImpl extends ExtendHibernateDaoSupport implements OrdersbookDao{

    /** 添加一条订单图书信息 */
    public void add(Ordersbook ordersbook){
        getHibernateTemplate().save(ordersbook);
    }

    /**根据 ordersId 获取该订单所有订单图书 */
    @SuppressWarnings("unchecked")
    public List<Ordersbook> allOrdersbookByOrders(int ordersId){
        return getHibernateTemplate().find("from Ordersbook as ordersbook where ordersbook.orders.ordersId= ?",ordersId);
    }
}
```

如下是 UserDao 组件的具体实现类 UserDaoImpl 的源代码：

```java
package iit.bookstore.dao.impl;
```

```java
import java.util.List;

import iit.bookstore.dao.UserDao;
import iit.bookstore.dao.base.ExtendHibernateDaoSupport;
import iit.bookstore.entity.User;

public class UserkDaoImpl extends ExtendHibernateDaoSupport implements UserDao {
    /** 获取用户信息 */
    public User findById(int userId) {
        return (User) getHibernateTemplate().get("entity.User", userId);
    }

    /** 通过username和password查询User对象,若有则返回User实例,否则null*/
    @SuppressWarnings("unchecked")
    public User findByNamePasswd(String userName, String userPassword) {
        List<User> userList = getHibernateTemplate().find("from User as user where user.userName = ? and user.userPassword = ?",
                new Object[]{userName,userPassword});
        if(userList != null && userList.size() > 0) {
            return userList.get(0);
        }
        return null;
    }

    /** 通过username查询User对象,若有则返回User实例,否则null*/
    @SuppressWarnings("unchecked")
    public User findByName(String userName) {
        List<User> userList = getHibernateTemplate().find( "from User as user where user.userName = ?",
                userName);
        if(userList != null && userList.size() > 0) {
            return userList.get(0);
        }
        return null;
    }

    /** 添加一条用户信息 */
    public int add(User user) {
        return (Integer)getHibernateTemplate().save(user);
    }

    /** 修改一条用户信息 */
    public void update(User user) {
        getHibernateTemplate().update(user);
    }
}
```

13.5.3 配置 DAO 组件

这些 DAO 组件已经定义和实现，但是为交给 Spring 框架进行管理，还需在 Spring 的配置文件进行配置，下面是 DAO 组件的配置文件代码：

如下是 BookDao 组件的具体实现类 BookDaoImpl 的源代码：

```xml
..............
<!--===================== 定义DAO组件======================= -->
<!-- 图书Dao -->
<bean id="bookDao" class="iit.bookstore.dao.impl.BookDaoImpl">
    <property name="sessionFactory" ref="sessionFactory"/>
</bean>
<!-- 特价图书Dao -->
<bean id="bargainDao" class="iit.bookstore.dao.impl.BargainDaoImpl">
    <property name="sessionFactory" ref="sessionFactory"/>
</bean>
<!-- 推荐图书Dao -->
<bean id="recommendedDao" class="iit.bookstore.dao.impl.RecommendedDaoImpl">
    <property name="sessionFactory" ref="sessionFactory"/>
</bean>
<!-- 图书类别Dao -->
<bean id="typeDao" class="iit.bookstore.dao.impl.TypeDaoImpl">
    <property name="sessionFactory" ref="sessionFactory"/>
</bean>
<!-- 管理员Dao -->
<bean id="managerDao" class="iit.bookstore.dao.impl.ManagerDaoImpl">
    <property name="sessionFactory" ref="sessionFactory"/>
</bean>
<!-- 普通用户Dao -->
<bean id="userDao" class="iit.bookstore.dao.impl.UserkDaoImpl">
    <property name="sessionFactory" ref="sessionFactory"/>
</bean>
<!-- 性别Dao -->
<bean id="sexDao" class="iit.bookstore.dao.impl.SexDaoImpl">
    <property name="sessionFactory" ref="sessionFactory"/>
</bean>
<!-- 订单Dao -->
<bean id="ordersDao" class="iit.bookstore.dao.impl.OrdersDaoImpl">
    <property name="sessionFactory" ref="sessionFactory"/>
</bean>
<!-- 订单图书Dao -->
<bean id="ordersbookDao" class="iit.bookstore.dao.impl.OrdersbookDaoImpl">
    <property name="sessionFactory" ref="sessionFactory"/>
</bean>
..............
```

13.6 创建业务层组件

13.6.1 业务逻辑组件接口的定义

本系统只使用了三个业务逻辑组件,分别为系统中最重要的三个功能模块提供业务逻辑实现:BookManager、OrdersManage 和 PersonManage 模块。这三个业务逻辑组件分别使用了不同的 DAO 组件,利用 DAO 组件的数据访问功能进行数据的相关操作。系统使用这三个业务逻辑组件将这些 DAO 组件封转在一起。

在这三个业务逻辑组件中也是采用面向接口的编程方式,下面是关于这些业务逻辑组件接口的源代码。

BookManager 接口的定义如下:

```java
package iit.bookstore.service;

import iit.bookstore.entity.Bargain;
import iit.bookstore.entity.Book;
import iit.bookstore.entity.Recommended;
import iit.bookstore.entity.Type;

import java.util.List;

public interface BookManage {

    /** 添加新的图书 */
    public int addBook(Book book);

    /** 修改图书 */
    public void updateBook(Book book);

    /** 根据图书ID查询该图书 */
    public Book findBook(int bookId);

    /** 查询销量最好的图书 */
    public List<Book> bestSellingBook(int pageNumber, int pageSize);

    /** 查询最新上架的图书 */
    public List<Book> latestBook(int pageNumber, int pageSize);

    /** 查询推荐图书ID */
    public List<Recommended> allRecommended(int pageNumber, int pageSize);

    /** 查询特价图书ID */
    public List<Bargain> allBargain(int pageNumber, int pageSize);

    /** 根据图书名称查询图书 */
    public List<Book> allBookByName(String bookName, int pageNumber,
        int pageSize);
```

第 13 章 Spring+Struts2+Hibernate 集成实例

```java
    /** 根据作者查询图书 */
    public List<Book> allBookByAuthor(String bookAuthor, int pageNumber,
            int pageSize);

    /** 根据出版社查询图书 */
    public List<Book> allBookByPress(String bookPress, int pageNumber,
            int pageSize);

    /** 根据类别ID来获取类别 */
    public Type findType(int typeId);

    /** 根据图书类别来查询图书 */
    public List<Book> allBookByType(int typeId, int pageNumber, int pageSize);

    /** 查询图书是否为特价图书 */
    public Bargain isBargain(int bookId);
}
```

OrdersManage 接口的定义如下：

```java
package iit.bookstore.service;

import iit.bookstore.entity.Orders;
import iit.bookstore.entity.Ordersbook;

import java.util.List;

public interface OrdersManage {

    /** 查询所有订单 */
    public List<Orders> allOrders(final int pageNumber, final int pageSize);

    /** 查询所有订单 */
    public List<Orders> allOrdersByDeal(final String isDeal,
            final int pageNumber, final int pageSize);

    /** 添加一个新的订单 */
    public int addOrders(Orders orders);

    /** 删除一个订单 */
    public void deleteOrders(int ordersId);

    /** 修改订单 */
    public void updateOrders(Orders orders);

    /** 根据ID查询订单 */
    public Orders findOrders(int ordersId);
```

```java
    /** 添加一条订单图书信息 */
    public void addOrdersbook(Ordersbook ordersbook);

    /** 根据userId获取该用户所有订单 */
    public List<Orders> allOrdersByUser(int userId, int pageNumber, int pageSize);

    /** 根据userId,订单处理状态获取该用户所有订单 */
    public List<Orders> allOrdersByUserDeal(int userId, String isDeal,
            int pageNumber, int pageSize);

    /** 根据ordersId获取该订单所有订单图书 */
    public List<Ordersbook> allOrdersbookByOrders(int ordersId);

}
```

PersonManage 接口的定义如下：

```java
package iit.bookstore.service;

import iit.bookstore.entity.Sex;
import iit.bookstore.entity.User;

public interface PersonManage {

    /**管理员登录验证,若通过则返回true, 否则返回false */
    public boolean checkManager(String managerName, String managerPassword);

    /** 获取用户信息 */
    public User findUser(int userId);

    /** 普通用户登录验证 ,若验证通过则返回User 实例, 否则返回null*/
    public User checkUser(String userName, String userPassword);

    /** 检查注册用户名是否已经存在 */
    public boolean isUserNameExist(String userName);

    /** 添加一条用户信息 */
    public int addUser(User user);

    /** 修改一条用户信息 */
    public void updateUserInfor(User user);

    /** 根据sexId查询Sex */
    public Sex findSex(int sexId);

}
```

13.6.2 实现业务逻辑组件

业务逻辑组件负责实现系统所需的业务方法，系统有多少个业务需求，业务逻辑组件就提供多少个对应的方法。本系统中的所有业务逻辑方法完全由业务逻辑组件负责实现。业务逻辑组件只负责业务逻辑上的变化，具体的数据持久化操作交给 DAO 层负责，因此业务逻辑组件是依赖于 DAO 组件的。下面是业务逻辑组件实现的源代码。

如下是 BookManager 组件的具体实现类 BookManagerImpl 的源代码：

```java
package iit.bookstore.service.impl;

import iit.bookstore.dao.BargainDao;
import iit.bookstore.dao.BookDao;
import iit.bookstore.dao.RecommendedDao;
import iit.bookstore.dao.TypeDao;
import iit.bookstore.entity.Bargain;
import iit.bookstore.entity.Book;
import iit.bookstore.entity.Recommended;
import iit.bookstore.entity.Type;
import iit.bookstore.service.BookManage;

import java.util.List;

public class BookManageImpl implements BookManage {
    /** Dao 组件注入 */
    private BookDao bookDao;
    private RecommendedDao recommendedDao;
    private BargainDao bargainDao;
    private TypeDao typeDao;

    public void setBookDao(BookDao bookDao) {
        this.bookDao = bookDao;
    }

    public void setRecommendedDao(RecommendedDao recommendedDao) {
        this.recommendedDao = recommendedDao;
    }

    public void setBargainDao(BargainDao bargainDao) {
        this.bargainDao = bargainDao;
    }

    public void setTypeDao(TypeDao typeDao) {
        this.typeDao = typeDao;
    }

    /** 添加新的图书 */
    public int addBook(Book book) {
        return bookDao.add(book);
    }
```

```java
/** 修改图书 */
public void updateBook(Book book) {
    bookDao.update(book);
}

/** 根据图书ID查询该图书 */
public Book findBook(int bookId) {
    return bookDao.findById(bookId);
}

/** 查询销量最好的图书 */
public List<Book> bestSellingBook(int pageNumber, int pageSize) {
    return bookDao.bestSellingBook(pageNumber, pageSize);
}

/** 查询最新上架的图书 */
public List<Book> latestBook(int pageNumber, int pageSize) {
    return bookDao.latestBook(pageNumber, pageSize);
}

/** 查询推荐图书ID */
public List<Recommended> allRecommended(int pageNumber, int pageSize) {
    return recommendedDao.findAll(pageNumber, pageSize);
}

/** 查询特价图书ID */
public List<Bargain> allBargain(int pageNumber, int pageSize) {
    return bargainDao.findByPage(pageNumber, pageSize);
}

/** 根据图书名称查询图书 */
public List<Book> allBookByName(String bookName, int pageNumber,
        int pageSize) {
    return bookDao.allBookByName(bookName, pageNumber, pageSize);
}

/** 根据作者查询图书 */
public List<Book> allBookByAuthor(String bookAuthor, int pageNumber,
        int pageSize) {
    return bookDao.allBookByAuthor(bookAuthor, pageNumber, pageSize);
}

/** 根据出版社查询图书 */
public List<Book> allBookByPress(String bookPress, int pageNumber,
        int pageSize) {
    return bookDao.allBookByPress(bookPress, pageNumber, pageSize);
}

/** 根据类别ID来获取类别 */
```

```java
    public Type findType(int typeId) {
        return typeDao.findById(typeId);
    }

    /** 根据图书类别来查询图书 */
    public List<Book> allBookByType(int typeId, int pageNumber, int pageSize) {
        return bookDao.allBookByType(typeId, pageNumber, pageSize);
    }

    /** 查询图书是否为特价图书 */
    public Bargain isBargain(int bookId) {
        return bargainDao.isBargain(bookId);
    }
}
```

如下是OrdersManage组件的具体实现类OrdersManageImpl的源代码：

```java
package iit.bookstore.service.impl;

import iit.bookstore.dao.OrdersDao;
import iit.bookstore.dao.OrdersbookDao;
import iit.bookstore.entity.Orders;
import iit.bookstore.entity.Ordersbook;
import iit.bookstore.service.OrdersManage;
import java.util.List;
public class OrdersManageImpl implements OrdersManage{
    /** Dao层组件的注入 */
    private OrdersDao ordersDao;
    private OrdersbookDao ordersbookDao;

    public void setOrdersDao(OrdersDao ordersDao) {
        this.ordersDao = ordersDao;
    }
    public void setOrdersbookDao(OrdersbookDao ordersbookDao) {
        this.ordersbookDao = ordersbookDao;
    }

    /** 查询所有订单 */
    public List<Orders> allOrders(final int pageNumber,final int pageSize){
        return ordersDao.findAll(pageNumber, pageSize);
    }

    /** 查询所有订单-根据订单状态 */
    public List<Orders> allOrdersByDeal(final String isDeal,final int pageNumber,final int pageSize){
        return ordersDao.allOrdersByDeal(isDeal, pageNumber, pageSize);
    }

    /** 添加一个新的订单 */
```

```java
    public int addOrders(Orders orders){
        return ordersDao.add(orders);
    }

    /** 删除一个订单 */
    public void deleteOrders(int ordersId){
        ordersDao.delete(ordersId);
    }

    /** 修改订单 */
    public void updateOrders(Orders orders){
        ordersDao.update(orders);
    }

    /** 根据ID查询订单 */
    public Orders findOrders(int ordersId){
        return ordersDao.findById(ordersId);
    }

    /** 添加一条订单图书信息 */
    public void addOrdersbook(Ordersbook ordersbook){
        ordersbookDao.add(ordersbook);
    }

    /** 根据userId获取该用户所有订单 */
    public List<Orders> allOrdersByUser(int userId,int pageNumber,int pageSize){
        return ordersDao.allOrdersByUser(userId, pageNumber, pageSize);
    }
    /** 根据userId,订单处理状态获取该用户所有订单 */
    public List<Orders> allOrdersByUserDeal(int userId,String isDeal,int pageNumber,int pageSize){
        return ordersDao.allOrdersByUserDeal(userId, isDeal, pageNumber, pageSize);
    }

    /** 根据ordersId获取该订单所有订单图书 */
    public List<Ordersbook> allOrdersbookByOrders(int ordersId){
        return ordersbookDao.allOrdersbookByOrders(ordersId);
    }
}
```

如下 PersonManage 组件的具体实现类 PersonManageImpl 的源代码:

```java
package iit.bookstore.service.impl;

import iit.bookstore.dao.ManagerDao;
import iit.bookstore.dao.SexDao;
import iit.bookstore.dao.UserDao;
import iit.bookstore.entity.Sex;
import iit.bookstore.entity.User;
```

```java
import iit.bookstore.service.PersonManage;

public class PersonManageImpl implements PersonManage {
    /** Dao组件的注入 */
    private UserDao userDao;
    private ManagerDao managerDao;
    private SexDao sexDao;

    public void setUserDao(UserDao userDao) {
        this.userDao = userDao;
    }

    public void setManagerDao(ManagerDao managerDao) {
        this.managerDao = managerDao;
    }

    public void setSexDao(SexDao sexDao) {
        this.sexDao = sexDao;
    }

    /**管理员登录验证,若通过则返回true，否则返回false */
    public boolean checkManager(String managerName,String managerPassword){
        return managerDao.checkManager(managerName, managerPassword);
    }

    /** 获取用户信息 */
    public User findUser(int userId){
        return userDao.findById(userId);
    }

    /** 普通用户登录验证 ,若验证通过则返回User 实例，否则返回null*/
    public User checkUser(String userName,String userPassword){
        return userDao.findByNamePasswd(userName, userPassword);
    }
    /** 检查注册用户名是否已经存在 */
    public boolean isUserNameExist(String userName){
        return userDao.findByName(userName)!=null;
    }
    /** 添加一条用户信息 */
    public int addUser(User user){
        return userDao.add(user);
    }
    /** 修改一条用户信息 */
    public void updateUserInfor(User user){
        userDao.update(user);
    }
    /** 根据sexId查询Sex */
    public Sex findSex(int sexId){
```

```
        return sexDao.findById(sexId);
    }
}
```

13.6.3 事务管理配置

与所有的 Java EE 项目类似，本系统的事务管理负责管理业务逻辑组件里面的业务逻辑方法，只有对业务逻辑方法添加事务管理才有实际的意义。

借助 Spring Schema 所提供的 tx、aop 两个命名空间的帮助，系统可以非常方便地为业务逻辑组件提供事务管理功能。下面是本应用中事务管理的配置代码：

```xml
...
<!-- 配置事务切面 Bean,指定事务管理器 -->
<tx:advice id="txAdvice" transaction-manager="transactionManager">
    <!-- 用于配置详细的事务语义 -->
    <tx:attributes>
        <!-- 所有以'get'开头的方法是 read-only 的 -->
        <tx:method name="get*" read-only="true"/>
        <!-- 其他方法使用默认的事务设置 -->
        <tx:method name="*"/>
    </tx:attributes>
</tx:advice>
<aop:config>
    <!-- 配置一个切入点，匹配 iit.bookstore.service.impl 包下
        所有以 Impl 结尾的类的所有方法的执行 -->
    <aop:pointcut id="iitPointcut"
        expression="execution(* iit.bookstore.service.impl.*Impl.*(..))"/>
    <!-- 指定在 txAdvice 切入点应用 txAdvice 事务切面 -->
    <aop:advisor advice-ref="txAdvice"
        pointcut-ref="iitPointcut"/>
</aop:config>
...
```

通过上面提供的配置代码，系统会自动为 iit.bookstore.service.impl 包下所有类的方法增加事务管理，这样的事务配置方式非常简洁，可以很好地简化过程及简化 Spring 文件的配置。

13.6.4 配置业务逻辑组件

这些业务逻辑组件已经定义和实现，但是为交给 Spring 框架进行管理，还需在 Spring 的配置文件进行配置，下面是业务逻辑组件的配置文件代码：

```xml
    ...
    <!--==================定义业务逻辑组件========================-->
    <!-- 图书管理 -->
    <bean id="bookManage" class="iit.bookstore.service.impl.BookManageImpl">
        <property name="bookDao" ref="bookDao"/>
        <property name="recommendedDao" ref="recommendedDao"/>
        <property name="bargainDao" ref="bargainDao"/>
```

```xml
            <property name="typeDao" ref="typeDao"/>
        </bean>
        <!-- 订单管理 -->
        <bean id="ordersManage" class="iit.bookstore.service.impl.OrdersManageImpl">
            <property name="ordersDao" ref="ordersDao"/>
            <property name="ordersbookDao" ref="ordersbookDao"/>
        </bean>
        <!-- 用户管理 -->
        <bean id="personManage" class="iit.bookstore.service.impl.PersonManageImpl">
            <property name="userDao" ref="userDao"/>
            <property name="managerDao" ref="managerDao"/>
            <property name="sexDao" ref="sexDao"/>
        </bean>
...
```

13.7 创建业务控制器

13.7.1 业务控制器的执行流程

业务控制器的处理流程如下：

（1）浏览器发送请求。

（2）中心处理器根据 struts.xml 文件查找对应的处理请求的 Action 类。

（3）WebWork 的拦截器链自动对请求应用通用功能，例如：WorkFlow、Validation 等功能。

（4）如果 Struts.xml 文件中配置 Method 参数，则调用 Method 参数对应的 Action 类中的 Method 方法，否则调用通用的 Execute 方法来处理用户请求。

（5）将 Action 类中的对应方法返回的结果响应给浏览器。

13.7.2 网上书店系统 Action 类分析

通过 LoginAction 里的 Action 类进行讲解，该类的源代码如下：

```java
package iit.bookstore.action;

import iit.bookstore.entity.User;
import iit.bookstore.service.PersonManage;

import java.util.Map;

import com.opensymphony.xwork2.ActionContext;
import com.opensymphony.xwork2.ActionSupport;


@SuppressWarnings("serial")
public class LoginAction extends ActionSupport{
```

```java
    //登录用户名
    private String userName;
    //密码
    private String userPassword;
    //业务逻辑组件
    private PersonManage personManage;
    public String getUserName() {
        return userName;
    }
    public void setUserName(String userName) {
        this.userName = userName;
    }
    public String getUserPassword() {
        return userPassword;
    }
    public void setUserPassword(String userPassword) {
        this.userPassword = userPassword;
    }
    public void setPersonManage(PersonManage personManage) {
        this.personManage = personManage;
    }


    //普通用户登录
    @SuppressWarnings({ "rawtypes", "unchecked" })
    public String loginCheck(){
        User user = personManage.checkUser(userName, userPassword);
        if(user != null){
            Map session = ActionContext.getContext().getSession();
            if(session.get("loginUser") == null){
                session.put("loginUser", user);
                return SUCCESS;
            }
        }
        return ERROR;
    }
    //管理员登录
    @SuppressWarnings({ "unchecked" })
    public String managerLoginCheck(){
        if( personManage.checkManager(userName, userPassword)){
ActionContext.getContext().getSession().put("managerLoginName", userName);
            return SUCCESS;
        }
        return ERROR;
    }
}
```

通过对输入信息的判断，并调用业务逻辑层组件实现对登录信息的验证，在控制层代码的信息量很少，主要是对流程判断和控制，具体的业务逻辑处理只需要调用业务逻辑层的相应方法进行处理即可。下面看在 struts.xml 中，如何定义这个 Action 类，代码如下：

```xml
..............
<!-- 普通用户登录action -->
<action name="loginAction" class="LoginAction" method="loginCheck">
    <result name="error">/login.jsp </result>
    <result name="success" type="redirectAction">
        <param name="actionName">mainAction</param>
    </result>
</action>
..............
```

从上面代码看出，成功时跳转到 mainAction 页面，失败时返回 login.jsp。注意 class 属性并没有给出完整的路径，而是在 spring 的 applicationContext.xml 配置文件中对类文件进行了定义，如下所示：

```xml
<!-- 用户登录action -->
<bean id="LoginAction" class="iit.bookstore.action.LoginAction" scope="prototype">
    <property name="personManage" ref="personManage" />
</bean>
```

所以，此时 spring 负责该具体实现类的依赖注入、创建和管理。通过这种方式实现 struts 与 spring 框架的无缝结合，便于开发应用程序过程中的解构，并提高代码的可读性。

由于该系统中 Action 类非常多，此处不再一一列举，只要把控制层的处理流程搞清楚，代码的理解是非常容易的。

13.8 创建视图 JSP 页面

13.8.1 用户注册界面

用户登录界面 enroll.jsp 代码如下：

```jsp
<%@ page language="java" pageEncoding="gb2312"%>
<%@taglib uri="/struts-tags" prefix="s"%>

<html>
    <head>
        <title>用户注册</title>
    </head>
    <body>
        <center>
            <jsp:include page="top.jsp"></jsp:include>
            <div id="enroll">
                <h1>用户注册</h1><br/>
                <s:form action="enrollAction">
                    <s:textfield label="用户名" name="userName"></s:textfield>
                    <s:password label="密码" name="userPassword"></s:password>
                    <s:password label="重复密码" name="userRePassword">
</s:password>
```

```
            <s:textfield label="邮箱" name="userEmail"></s:textfield>
            <s:submit value="注册"></s:submit>
        </s:form>
      </div>
      <jsp:include page="bottom.jsp"></jsp:include>
    </center>
  </body>
</html>
```

13.8.2 用户登录界面

用户登录界面 login.jsp 代码如下：

```
<%@ page language="java" pageEncoding="utf-8"%>
<%@taglib uri="/struts-tags" prefix="s"%>
<html>
  <head>
     <title>用户登录</title>
  </head>
  <body>
  <center>
    <jsp:include page="top.jsp"></jsp:include>
    <div id="login">
      <h1>用户登录</h1><br/><!--  -->
      <s:form action="loginAction">
          <s:textfield label="用户名" name="userName"></s:textfield>
          <s:password label="密码" name="userPassword"></s:password>
          <s:submit value="登录"></s:submit>
          <s:fielderror></s:fielderror>
      </s:form>
    </div>
    <jsp:include page="bottom.jsp"></jsp:include>
  </center>
  </body>
</html>
```

13.8.3 用户信息修改界面

用户信息修改界面 personalInformation.jsp 代码如下：

```
<%@ page language="java" import="java.util.*" pageEncoding="utf-8"%>
<%@taglib uri="/struts-tags" prefix="s"%>
<!DOCTYPE HTML PUBLIC "-//W3C//DTD HTML 4.01 Transitional//EN">
<html>
  <head>

    <title>个人信息</title>
    <link rel="stylesheet" type="text/css" href="css/styles.css">
  </head>
```

第 13 章 Spring+Struts2+Hibernate 集成实例

```jsp
<body>
    <center>
        <jsp:include page="top.jsp"></jsp:include>
        <div id="personalInformation">
        <s:if test="%{#session.loginUser != null}">
            <h1>个人信息</h1><br/>
            <s:form action="userInforAction" method="post">
                <s:label label="用户名" name="userName" value="%{#session.loginUser.userName}"></s:label>
                <s:textfield label="昵称" name="userNickname" value="%{#session.loginUser.userNickname}"></s:textfield>
                <s:select label="性别" name="sexId" list="#{'1':'男','2':'女','3':'未知'}" value="%{#session.loginUser.sexId}"></s:select>
                <s:textfield label="邮箱" name="userEmail" value="%{#session.loginUser.userEmail}"></s:textfield>
                <s:textfield label="电话" name="userPhone" value="%{#session.loginUser.userPhone}"></s:textfield>
                <s:textfield label="地址" name="userAddress" value="%{#session.loginUser.userAddress}"></s:textfield>
                <s:textarea label="备注" name="userRemark" value="%{#session.loginUser.userRemark}"></s:textarea>
                <s:submit value="修改"></s:submit>
            </s:form>
            <a href="updatePassword.jsp">修改密码</a>
        </s:if>
        <s:else>
            <jsp:forward page="firstPage.jsp"></jsp:forward>
        </s:else>
        </div>
        <jsp:include page="bottom.jsp"></jsp:include>
    </center>
</body>
</html>
```

13.8.4 系统首页界面

系统首页界面 allBook.jsp 代码如下：

```jsp
<%@ page language="java" pageEncoding="gb2312"%>

<%@taglib prefix="s" uri="/struts-tags"%>

<!DOCTYPE HTML PUBLIC "-//W3C//DTD HTML 4.01 Transitional//EN">
<html>
  <head>
    <link rel="stylesheet" type="text/css" href="css/styles.css">
  </head>

  <body>
```

```html
        <center>
            <div id="allBook">
                <ul class="allBookUl">
                    <li class="allBookHead">
                        最新上架<a class="more" href="oneTypeAction.action?searchType=bookStatus&searchDescribe=latest">更多..</a>
                    </li>
                    <s:iterator value="latestBook" var="book" status="st">
                        <li class="allBookPicture">
                            <a href="singleBookAction.action?bookId=<s:property value="#book.bookId" />"><img src='upload/<s:property value="#book.bookPicture" />'/></a>
                        </li>
                        <li class="allBookInfor">
                            <a class="bookName" href="singleBookAction.action?bookId=<s:property value="bookId" />"><s:property value="#book.bookName" /></a><br/><br/>
                            作者：<s:property value="#book.bookAuthor"/><br/><br/>
                            出版社：<s:property value="#book.bookPress"/><br/><br/>
                            类别：<a class="aboutBook" href="oneTypeAction.action?searchType=bookType&searchDescribe=<s:property value="#book.type.typeId"/>"><s:property value="type.typeName"/></a><br/><br/>
                        </li>
                    </s:iterator>
                </ul>
                <ul class="allBookUl">
                    <li class="allBookHead">
                        特价热销<a class="more" href="oneTypeAction.action?searchType=bookStatus&searchDescribe=bargain">更多..</a>
                    </li>
                    <s:iterator value="bargainBook">
                        <li class="allBookPicture">
                            <a href="singleBookAction.action?bookId=<s:property value="book.bookId" />"><img src='upload/<s:property value="book.bookPicture" />'/></a>
                        </li>
                        <li class="allBookInfor">
                            <a class="bookName" href="singleBookAction.action?bookId=<s:property value="book.bookId" />"><s:property value="book.bookName" /></a><br/><br/>
                            作者:<s:property value="book.bookAuthor"/><br/><br/>
                            出版社：<s:property value="book.bookPress"/><br/><br/>
                            类别：<a class="aboutBook" href="oneTypeAction.action?searchType=bookType&searchDescribe=<s:property value="book.type.typeId"/>"><s:property value="book.type.typeName"/></a><br/><br/>
                        </li>
                    </s:iterator>
                </ul>
                <ul class="allBookUl">
                    <li class="allBookHead">
```

```
                         精品推荐<a class="more" href="oneTypeAction.action?
searchType=bookStatus&searchDescribe=recommended">更多..</a>
                    </li>
                    <s:iterator value="recommendedBook">
                        <li class="allBookPicture">
                            <a href="singleBookAction.action?bookId=<s:property
value="book.bookId" />"><img src='upload/<s:property value="book.bookPicture"
/>'/></a>
                        </li>
                        <li class="allBookInfor">
                            <a class="bookName" href="singleBookAction.action?
bookId=<s:property value="book.bookId" />"><s:property value="book.bookName"
/></a><br/><br/>
                            作者:<s:property value="book.bookAuthor"/><br/><br/>
                            出版社: <s:property value="book.bookPress"/><br/><br/>
                            类别: <a class="aboutBook" href="oneTypeAction.action?
searchType=bookType&searchDescribe=<s:property value="book.type.typeId"/>">
<s:property value="book.type.typeName"/></a><br/><br/>
                        </li>
                    </s:iterator>
                </ul>
            </div>
        </center>
    </body>
</html>
```

13.8.5　显示图书详细信息界面

显示图书详细信息界面 singleBook.jsp 代码如下：

```
<%@ page language="java" pageEncoding="utf-8"%>
<%@taglib uri="/struts-tags" prefix="s"%>

<!DOCTYPE HTML PUBLIC "-//W3C//DTD HTML 4.01 Transitional//EN">
  <head>
    <link rel="stylesheet" type="text/css" href="css/styles.css">
  </head>

  <body>
    <center>
        <jsp:include page="top.jsp"></jsp:include>
        <div id="singleBook">
            <div id="left">
                <jsp:include page="left.jsp"></jsp:include>
            </div>
            <div id="bookInfor">
                <ul class="singleBookUl">
                    <li class="singleBookName">
                        <s:property value="singleBook.bookName"/>
                    </li>
```

```html
                    <li class="singleBookPicture">
                        <img src='upload/<s:property value="singleBook.bookPicture"/>' />
                    </li>
                    <li class="singleBookInfor">
                        作者：<a class="aboutBook" href="oneTypeAction.action?searchType=bookAuthor&searchDescribe=<s:property value="singleBook.bookAuthor"/>"><s:property value="singleBook.bookAuthor"/></a><br/>
                        出版社：<a class="aboutBook" href="oneTypeAction.action?searchType=bookPress&searchDescribe=<s:property value="singleBook.bookPress"/>"><s:property value="singleBook.bookPress"/></a><br/>
                        类别：<a class="aboutBook" href="oneTypeAction.action?searchType=bookType&searchDescribe=<s:property value="singleBook.type.typeId"/>"><s:property value="singleBook.type.typeName"/></a><br/>
                        上架时间：<s:date name="singleBook.bookShelveTime" format="yyyy-MM-dd HH:mm:ss"/><br/>
                        图书简介：<s:property value="singleBook.bookRemark"/><br/>
                        原价：<font style="text-decoration: line-through; color:red;"><s:property value="singleBook.bookPrice"/></font> 元    现价：<s:property value="singleBook.bookNewPrice"/> 元<br/>
                        <a href='shoppingCartAction.action?bookId=<s:property value="singleBook.bookId"/>'>放入购物车</a>
                    </li>
                </ul>
            </div>
        </div>
        <jsp:include page="bottom.jsp"></jsp:include>
    </center>
</body>
```

13.8.6 购物车界面

购物车界面 shoppingCartr.jsp 代码如下：

```jsp
<%@ page language="java" import="java.util.*" pageEncoding="utf-8"%>
<%@taglib uri="/struts-tags" prefix="s"%>
<!DOCTYPE HTML PUBLIC "-//W3C//DTD HTML 4.01 Transitional//EN">
<html>
  <head>
    <title>购物车</title>
    <link rel="stylesheet" type="text/css" href="css/styles.css">
  </head>

  <body>
    <center>
        <jsp:include page="top.jsp"></jsp:include>
        <div id="shoppingCart">
            <div id="left">
                <jsp:include page="left.jsp"></jsp:include>
```

```html
            </div>
            <div id="shoppingBook" style="padding-top: 20px;padding-left: 8px;">
                <ul class="shoppingBookUl">
                    <li class="shoppingBookHead">我的购物车</li>
                </ul>
                <ul class="shoppingBookUl" style="background-color: yellow;">
                    <li class="sequence">序列</li>
                    <li class="bookName">图书名称</li>
                    <li class="bookPrice">图书价格</li>
                    <li class="bookAmount" style="padding-top: 5px;">购买数量</li>
                    <li class="delete" style="padding-top: 5px;">删除图书</li>
                </ul>
                <s:if test="%{#session.shoppingCart != null}">
                    <s:iterator value="#session.shoppingCart.items" status="st">
                        <ul class="shoppingBookUl">
                            <li class="sequence">
                                <s:property value="#st.getIndex()+1"/>
                            </li>
                            <li class="bookName"><a class="bookName" href="singleBookAction.action?bookId=<s:property value="book.bookId"/>"><s:property value="book.bookName"/></a></li>
                            <li class="bookPrice"><s:property value="book.bookPrice"/> 元</li>
                            <li class="bookAmount">
                                <input type="text" style="width:80px;" id="bookAmount<s:property value="book.bookId"/>" value='<s:property value="amount"/>'/>
                                <input type="button" value="修改" onclick="updateBookAmount('<s:property value="book.bookId"/>')">
                            </li>
                            <li class="delete">
                                <input type="button" value="删除" onclick="deleteBook('<s:property value="book.bookId"/>')">
                            </li>
                        </ul>
                    </s:iterator>
                    <ul class="shoppingBookUl">
                        <li class="shoppingBookHead">
                        <s:if test="%{#session.shoppingCart.size() > 0}">
                            <input type="button" value="确定购买" onclick="addOrders()">
                        </s:if>
                            总计金额：<s:property value="#session.shoppingCart.totalMoney"/> 元
                        </li>
                    </ul>
                </s:if>
            </div>
```

```
        </div>
        <jsp:include page="bottom.jsp"></jsp:include>
    </center>
</body>
<SCRIPT type="text/javascript">
<!--
    function updateBookAmount(bookId){
        var pattern = /^[1-9][0-9]{0,}$/;
        var bookAmount = document.getElementById("bookAmount"+bookId).value;
        if(pattern.test(bookAmount)){
            location.href = "updateCartAction.action?updateType=update&bookId="+bookId+"&bookAmount="+bookAmount;
        }
    }
    function deleteBook(bookId){
        if(confirm("确定要删除吗？")){
            location.href = "updateCartAction.action?updateType=delete&bookId="+bookId;
        }
    }
    function addOrders(){
        if(confirm("确定要购买吗？")){
            location.href = "ordersManageAction.action?updateType=add";
        }
    }
//-->
</SCRIPT>
</html>
```

13.8.7 显示用户订单列表界面

显示用户订单列表界面 allOrders.jsp 代码如下：

```
<%@ page language="java" import="java.util.*" pageEncoding="utf-8"%>
<%@taglib prefix="s" uri="/struts-tags"%>
<!DOCTYPE HTML PUBLIC "-//W3C//DTD HTML 4.01 Transitional//EN">
<html>
  <head>
    <title>我的订单</title>

    <meta http-equiv="pragma" content="no-cache">
    <meta http-equiv="cache-control" content="no-cache">
    <meta http-equiv="expires" content="0">
    <meta http-equiv="keywords" content="keyword1,keyword2,keyword3">
    <meta http-equiv="description" content="This is my page">
    <link rel="stylesheet" type="text/css" href="css/styles.css">

  </head>

<body>
    <center>
```

```jsp
            <jsp:include page="top.jsp"></jsp:include>
            <div id="orders">
                <div id="left">
                    <jsp:include page="left.jsp"></jsp:include>
                </div>
                <div id="ordersInfor" style="padding-top: 20px;padding-left: 8px;">
                    <ul class="singleOrders" style="background-color: teal;">
                        <li style="width: 700px;padding-top: 5px;padding-left: 10px;text-align: left;">
                            <a class="aboutOrders" href="ordersAction.action?searchType=all">所有订单</a>    
                            <a class="aboutOrders" href="ordersAction.action?searchType=isDeal">已处理订单</a>    
                            <a class="aboutOrders" href="ordersAction.action?searchType=isNotDeal">未处理订单</a>
                        </li>
                    </ul>
                    <ul class="singleOrders" style="background-color: yellow;">
                        <li class="sequence">序列</li>
                        <li class="ordersNumber">订单编号</li>
                        <li class="ordersTime">订单日期</li>
                        <li class="isDeal">处理状态</li>
                        <li class="deleteOrders" style="padding-top: 5px;">删除订单</li>
                    </ul>
                    <s:iterator value="#request.allOrdersByUser" status="st">
                        <ul class="singleOrders">
                            <li class="sequence"><s:property value="#st.getIndex()+#request.sequence+1"/></li>
                            <li class="ordersNumber">
                                <a class="aboutBook" href="singleOrdersAction.action?ordersId=<s:property value="ordersId"/>"><s:property value="ordersNumber"/></a>
                            </li>
                            <li class="ordersTime"><s:date name="ordersTime" format="yyyy-MM-dd HH:mm:ss"/></li>
                            <li class="isDeal">
                                <s:if test='%{isDeal =="0"}'>
                                    <font style="color: green;">未处理</font>
                                </s:if>
                                <s:else>
                                    <font style="color: red;">已处理</font>
                                </s:else>
                            </li>
                            <li class="deleteOrders">
                                <input type="button" value="删除" onclick="deleteOrders('<s:property value="ordersId"/>')" />
                            </li>
                        </ul>
                    </s:iterator>
                </div>
```

```
            </div>
            <jsp:include page="bottom.jsp"></jsp:include>
        </center>
    </body>
    <SCRIPT type="text/javascript">
    <!--
        function deleteOrders(ordersId){
            if(confirm("确定要订单吗？")){
                location.href="ordersManageAction.action?updateType=delete&ordersId="+ordersId;
            }
        }
    //-->
    </SCRIPT>
</html>
```

13.8.8 添加图书界面

管理员添加图书界面 addBook.jsp 代码如下：

```
<%@ page language="java" import="java.util.*" pageEncoding="utf-8"%>
<%@taglib prefix="s" uri="/struts-tags"%>
<!DOCTYPE HTML PUBLIC "-//W3C//DTD HTML 4.01 Transitional//EN">
<html>
    <head>
        <title>添加图书</title>
        <link rel="stylesheet" type="text/css" href="css/style.css">

    </head>

    <body>
        <center>
            <jsp:include page="top.jsp"></jsp:include>
            <div id="managePage">
                <div id="manageLeft">
                    <jsp:include page="manageLeft.jsp"></jsp:include>
                </div>
                <div id="addBook">
                    <h1>添加新书</h1><br/>
                    <s:form action="bookAction" method="post" enctype="multipart/form-data">
                        <s:textfield label="名称" name="bookName"></s:textfield>
                        <s:textfield label="作者" name="bookAuthor"></s:textfield>
                        <s:textfield label="出版社" name="bookPress"></s:textfield>
                        <s:file label="图片" name="doc"></s:file>
                        <s:select label="类别" name="typeId" list="#{'1':'文学','2':'历史','3':'天文','4':'地理','5':'其他'}"></s:select>
                        <s:textfield label="价格" name="bookPrice"></s:textfield>
                        <s:textfield label="数量" name="bookAmount"></s:textfield>
                        <s:textarea label="简介" name="bookRemark"></s:textarea>
                        <s:submit value="添加"></s:submit>
                    </s:form>
                </div>
            </div>
```

```
            <jsp:include page="bottom.jsp"></jsp:include>
        </center>
    </body>
</html>
```

13.9 运行网上书店系统

13.9.1 系统前台界面

把该网上书店系统部署到 Tomcat 服务器上，然后输入网址：http://localhost:8080/bookStore，即可看到该系统的运行效果，如图 13-3～图 13-7 所示。

图 13-3 系统首页界面

图 13-4 用户注册界面

综合篇

图 13-5　用户信息修改界面

图 13-6　购物车界面

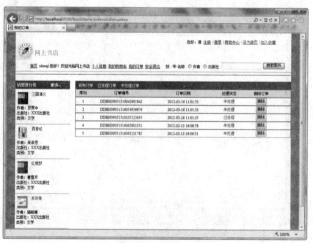

图 13-7　用户订单查看界面

13.9.2 系统后台界面

系统后台界面如图 13-8～图 13-11 所示。

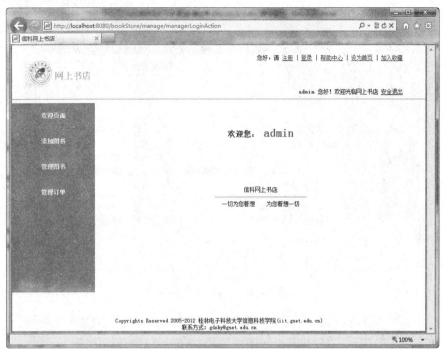

图 13-8 后台管理员欢迎界面

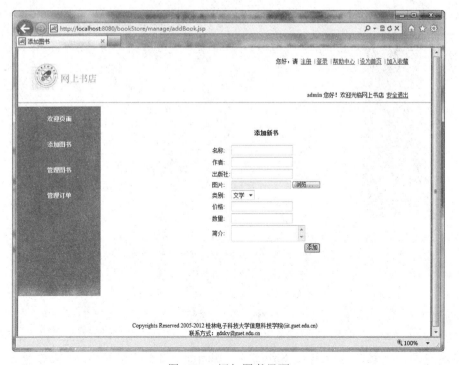

图 13-9 添加图书界面

图 13-10　图书管理界面

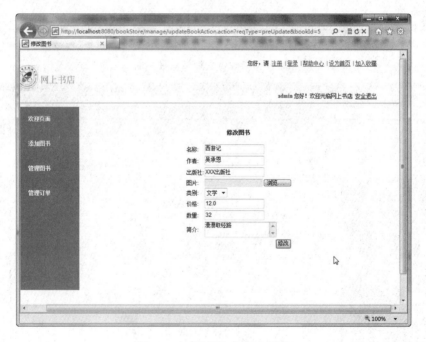

图 13-11　图书信息修改界面

小　　结

本章建立了一个基于 Spring、Struts2 和 Hibernate 三个框架的网上书店系统。在这个

第 13 章 Spring+Struts2+Hibernate 集成实例

网上书店系统中,详细介绍了一个完整的 Java EE 项目。读者学习本章节内容可能会有些难度,这涉及三个框架中的内容。一旦读者掌握了本章案例的开发思路,就会会对 Spring、Struts2 和 Hibernate 这三个框架有更深一层的理解,也会对实际 Java EE 项目开发有更高层次的认识和掌握。通过本章学习,读者将熟悉 Spring+Struts2+Hibernate 的框架模式的 Web 开发方法。

习　题

在本章网上书店系统的代码的基础上,对网上书店系统的功能进行扩展,使程序的功能更加完善。

附录 A 常见数据类型转换

```java
package com.common;
import java.math.BigDecimal;
import java.text.ParseException;
import java.text.SimpleDateFormat;
import java.util.Date;
/**
 * 数据转换类
 */
public class DataConverter {
    /**
     * 通过汇率将人民币转换为美元
     * @param rmb
     * @return
     */
    public static BigDecimal toUSD(String rmb){
        return toUSD(toBigDecimal(rmb));
    }

    /**
     * 通过汇率将人民币转换为美元
     * @param rmb
     * @return
     */
    public static BigDecimal toUSD(Object rmb){
        return toUSD(rmb.toString());
    }

    /**
     * 通过汇率将人民币转换为美元
     * @param rmb
     * @return
     */
    public static BigDecimal toUSD(BigDecimal rmb) {
        BigDecimal rate = new BigDecimal(SiteConfig.getInstance().getExchageRate());
        if(rate.equals(BigDecimal.ZERO) || rmb.equals(BigDecimal.ZERO)){
            return rmb;
        }
        try {
            BigDecimal value = rmb.divide(rate, 10, BigDecimal.ROUND_HALF_UP);
            return value.setScale(0, BigDecimal.ROUND_HALF_UP);
```

```java
        } catch (Exception e) {
            e.printStackTrace();
            return BigDecimal.ZERO;
        }
    }

    /**
     * 将日期格式转换为字符串
     * @param date
     * @return
     */
    public static String dataToString(Date date){
        return dataToString(date, "yyyy-MM-dd HH:mm:ss");
    }

    /**
     * 将日期格式转换为字符串
     * @param date - 日期
     * @param formatType - 格式化方式
     * @return - 字符串
     */
    public static String dataToString(Date date, String formatType){
        if(date == null){
            date = Utility.getNowDateTime();
        }
        SimpleDateFormat formatter = new SimpleDateFormat(formatType);
        return formatter.format(date);
    }

    /**
     * 将字符串转成日期(yyyy-MM-dd HH:mm:ss)
     * @param input - 日期字符串
     * @return
     */
    public static Date toDate(String input){
        return toDate(input, "yyyy-MM-dd HH:mm:ss");
    }

    /**
     * 将字符串转成日期
     * @param input - 日期字符串
     * @param formatType - 格式化类型,如: yyyy-MM-dd HH:mm:ss
     * @return 日期类型,当出现异常时返回当前日期
     */
    public static Date toDate(String input, String formatType){
        SimpleDateFormat format = new SimpleDateFormat(formatType);
        Date dt = new Date();
        if(DataValidator.isNullOrEmpty(input)){
            return dt;
        }
        try {
            dt = format.parse(input);
        } catch (ParseException e) {
        }
        return dt;
    }

    /**
```

```java
 * 将字符串转成短日期格式 yyyy-MM-dd
 * @param input - 日期字符串
 * @return 日期类型，当出现异常时返回当前日期
 */
public static java.util.Date toShortDate(String input){
    return toDate(input, "yyyy-MM-dd");
}

/**
 * 将字符串转成长日期格式 yyyy-MM-dd HH:mm:ss
 * @param input - 日期字符串
 * @return 日期类型，当出现异常时返回当前日期
 */
public static java.util.Date toFullDate(String input){
    return toDate(input, "yyyy-MM-dd HH:mm:ss");
}

/**
 * 将字符串转成整型
 * @param input - 要转换的字符串
 * @return 整数，出现异常则返回 0
 */
public static int toInt(String input){
    try{
        return Integer.parseInt(input);
    }catch(Exception e){
        return 0;
    }
}

/**
 * 将双精度转成整型
 * @param input
 * @return
 */
public static int toInt(double input){
    BigDecimal bd = new BigDecimal(input);
    return bd.setScale(0, BigDecimal.ROUND_HALF_UP).intValue();
}

/**
 * 将字符串转成双精度型
 * @param input - 要转换的字符串
 * @return 双精度数，出现异常则返回 0
 */
public static double toDouble(String input){
    try{
        return Double.parseDouble(input);
    }catch(Exception e){
        return 0;
    }
}

public static BigDecimal toBigDecimal(Object input){
    return toBigDecimal(input.toString());
}

public static BigDecimal toBigDecimal(String input){
```

```java
    try{
        BigDecimal bd = new BigDecimal(input);
        return bd;
    }catch(Exception e){
        return BigDecimal.ZERO;
    }
}

/**
 * 将字符串转成浮点型
 * @param input - 要转换的字符串
 * @return 浮点数，出现异常则返回 0
 */
public static float toFloat(String input){
    try{
        return Float.parseFloat(input);
    }catch(Exception e){
        return 0;
    }
}


/**
 * 将字符串转成长整型
 * @param input - 要转换的字符串
 * @return 长整型数，出现异常则返回 0
 */
public static long toLong(String input){
    try{
        return Long.parseLong(input);
    }catch(Exception e){
        return 0;
    }
}

public static short toShort(String input){
    try{
        return Short.parseShort(input);
    }catch(Exception e){
        return 0;
    }
}

/**
 * 将字符串转成布尔型
 * @param input - 要转换的字符串
 * @return 布尔值，出现异常则返回 false
 */
public static boolean toBoolean(String input){
    if(DataValidator.isNullOrEmpty(input)) return false;
    if(input.equals("1") || input.equals("on")) input = "true";
    if(input.equals("0") || input.equals("off")) input = "false";
    try{
        return Boolean.parseBoolean(input);
    }catch(Exception e){
        return false;
    }
}
}
```

参 考 文 献

[1] 刘晓华，张健，周慧贞．JSP 应用开发详解[M]．3 版．北京：电子工业出版社，2007．

[2] 刘中兵．开发者突击：Java Web 主流框架整合开发：J2EE+Struts+Hibernate+Spring[M]．北京：电子工业出版社，2011．

[3] 罗玉玲．J2EE 应用开发详解[M]．北京：电子工业出版社，2009．

[4] 杨少波．J2EE Web 核心技术：Web 组件与框架开发技术[M]．北京：清华大学出版社，2011．